LIFE'S INTRINSIC VALUE

LIFE'S INTRINSIC VALUE

Science, Ethics, and Nature

NICHOLAS AGAR

Columbia University Press • New York

COLUMBIA UNIVERSITY PRESS
PUBLISHERS SINCE 1893
NEW YORK   CHICHESTER, WEST SUSSEX

Library of Congress Cataloging-in-Publication Data

Agar, Nicholas.
   Life's intrinsic value ; science, ethics, and nature / Nicholas Agar.
      p.   cm.
   Includes bibliographical references
   ISBN: 978-0-231-11787-6 (pbk) -- ISBN: 978-0-231-11786-9 (cloth)
      1. Environmental ethics   2. Philosophy of nature.   I. Title.

GE42 .A38 2001
179'.1—dc21   00-047540

Contents

☙

Acknowledgments

☙

I have incurred many debts in the writing of this book. First, I should thank
Holly Hodder and Jonathan Slutsky, editors at Columbia University Press,
for shepherding the book through to production and Anthony Chiffolo,
copyeditor, for his hunting out of stylistic and argumentative infelicities.

I received specifically philosophical feedback from a number of sources.
Mostly anonymous referees for the press provided useful comments, both
sympathetic and combative. Various conference and department seminar
audiences throughout Australasia have heard and provided invaluable criticisms on sections of the book. I should also thank my colleagues here in
the Victoria University of Wellington Philosophy Department for years of
formal and informal feedback on the issues discussed in this book. Last, I
will single out for individual mention Ruth Avery, Hugh Clapin, Jack
Elder, Tony Fielding, Mike Hurst, Timothy Irwin, Edwin Mares, Alison
Munro, Denis Poole, Sarah Sandley, Kim Sterelny, and Gary Varner.

PREFACE

✺

In this book I aim to show that individual living things are intrinsically valuable and to found an environmental ethic on this value. The search for intrinsic value in nature will be familiar to readers already acquainted with the ethics of the environment. Demonstrating that environmental individuals or collections have intrinsic value seems the most straightforward way of showing that they matter morally. Yet in spite of the worthy efforts of such writers as J. Baird Callicott, Holmes Rolston III, and Arne Naess, we are some distance from a workable account of such value. The reason is not hard to see. Received ethical wisdom uses the notion of intrinsic value to indicate the great moral importance of humans. Stories about such value frequently latch onto perceived contrasts between humans and natural things. Rational humans are held to be ends, while nonrational termite mounds can only ever be means. We think of humans as experiencers of pleasure and suffering, but of kauri trees as mere instruments to these morally important experiences in other beings. Defenders of the supposed exclusive moral significance of humans can rely not only upon the inertia of commonsense moral belief but also upon centuries of philosophical speculation about the nature of value.

The life-centered or biocentric ethic requires some agent of moral transformation. I cast the sciences of the environment in this role. I show that science enables us to define a concept fit to spread intrinsic value beyond humanity. Just as in past centuries science dislodged humans from the center of the physical universe, it now challenges our claim to exclusive occupancy of the center of the moral universe.

How is science to effect this transformation? According to received philosophical wisdom, scientific theories simply do not provide the right variety of information to bring about fundamental changes to moral views.

I aim to refute this wisdom. The arguments of this book constitute a bridge between the things in the natural world that scientists undertake to describe and the psychological concepts that determine what types of things are intrinsically valuable. I introduce a psychological or "representational" conception of what it is for something to be alive, to expose the value of individual organisms ranging from the very simple to the very complex. The moral considerability of the representationally alive is defended within the context of broader views about what categories of things psychological terms refer to and the revisionary impact of science on these categories. A value-endowing psychological concept does not map straightforwardly onto any single pattern of objective sameness in nature. Rather, scientists discover a range of more or less perfect deservers of psychological names. Even the less perfect deservers are candidates for a degree of moral consideration. My account is not revisionist in that it does not seek to challenge the idea that to be valuable is to be in some important way psychological. It is revisionist in its use of science and the philosophy of language to extend the psychological far beyond the human.

My final task is to show that this ethic is more than an idle intellectual curiosity. The principles of the biocentric ethic are within reach of human beings.

LIFE'S INTRINSIC VALUE

The Psychological View of Intrinsic Value

LIFE ON MARS, LIFE ON EARTH

If the scientists at the NASA Johnson Space Center are to be believed, Martians finally came to earth not in a flying saucer but in a potato-sized piece of rock.[1] Some facts about meteorite ALH84001 are fairly certain. Formed around 4.5 billion years ago, it is easily the most ancient chunk of Mars to have been found on Earth. Perhaps a billion years after its creation, the rock suffered repeated meteoritic collisions, and water flowed through the resulting fractures. This allowed the formation of numerous tiny carbonate globules. The rock stayed put until around 16 million years ago, when it was bashed off the Martian surface, and after a few million years in space it landed in Antarctica, where it was discovered in 1984.

The carbonate globules that are the cause of ALH84001's celebrity could be the remains of Martian microbes, the fossils of contaminating Antarctic bacteria, or perhaps the never-alive products of geologic processes. Members of the NASA team give a range of reasons for thinking that they were once Martian microorganisms. First of all, the globules contain the organic molecules that are the building blocks of all life on Earth. Though the composition of carbon isotopes strongly suggests that the globules were formed from constituents originating in Mars's atmosphere, their shape and internal structure closely resemble those of confirmed Earthly microfossils. Like bacteria here, the occupants of ALH84001 are associated with iron sulfides and oxides. Mars may have witnessed climactic extremes over the past billions of years, but the globules appear to have formed at temperatures below 100 degrees Centigrade, within the range considered tolerable by bacteria.

This could-be Martian life has generated heated debate. Doubtless, continued intense scrutiny will someday allow us to file the nanobacteria of ALH84001 either under "scientific finds of the century" or together with cold fusion and liquefied-vegetable cures for cancer. Imagine for a while that the claims are confirmed. On Earth, organisms of similar size and simplicity evolved into creatures ranging from amoebas to wetas and coelacanths to gigantosaurs and blue whales. Could there be living descendants of these ancient Martian nanobacteria? Hopes were certainly once high that all manner of living things might today inhabit Mars. Nineteenth-century observers supposed that the lines visible on the surface of the planet were irrigation channels and that the periodic color changes were caused by extravagant summer vegetation growth forced into dormancy by the extreme winter. The 1965 flyby of *Mariner 4* dashed these hopes; twenty-two photographs revealed a surface scarred by craters and channels but no sign of irrigation or even of flowing water. Over ten years later the *Viking Landers* found no trace of life in Martian soil. It is remotely possible that surviving Martian life forms remain in underground hydrothermal springs, but today most of the interest centers on Mars's warmer, wetter distant past.[2]

Although microbes that last replicated and metabolized some 3.5 billion years ago are hardly about to invade this planet and conquer our civilization, the impact of even long-extinct Martian life forms would be considerable. One question concerns the origin of life on our own planet. Orthodox opinion holds that the oxygen-starved seas of an ancient Earth generated the first self-replicating structures out of relatively simple molecular and atomic constituents. Dissident scientists counter that this planet was seeded from space with amino acids that then combined to form the first living things.[3] If the nanobacteria of Earth and Mars share the same relatively complex building blocks, then an extraterrestrial origin for life may be more likely. The confirmed discovery of once-living things on our nearest neighbor would surely prompt us to ask how many other planets in the universe might be staging their own evolutionary dramas.

The existence of Martian life would have an impact far beyond our scientific picture of the universe. If individual living things are intrinsically valuable—and the aim of this book is to so demonstrate—then ALH84001 would move Mars from being a place that throughout its existence has been devoid of this most important brand of moral value to being somewhere that has had inhabitants fit to appear in the universal catalog of intrinsically valuable things. If the drying of the riverbeds really did destroy the last Martian microorganism, then this event would have been a moral disaster, in comparative terms, much more massive than the Cretaceous die-off or global nuclear war.

Any Martian revelations notwithstanding, it is probable that the value described by the life ethic is very thinly spread throughout the universe. Most supernovas or asteroid impacts will make fairly significant changes to the way natural properties are arranged in some solar system or on some planet, but only rarely will they disturb the universal distribution of intrinsic value.

This brings me to the first major task of this book. The life ethic will have the most profound and immediate implications for our view of our own planet. The intrinsic value associated with life forms the foundation of an environmental ethic, enabling us to recognize nature's moral importance.

The idea that nature is morally valuable should seem pretty familiar by now. After all, the environmental movement has been in full swing for more than thirty years—long enough for many people to assent unthinkingly to the claims that species ought to be preserved and that a world with thriving ecosystems is better than a world without them. Is the defense of such apparent moral truisms worth the effort, or does it reflect nothing more than a philosopher's obsession with justifying the obvious?

There is considerably more at stake than philosophical scruples. A thoroughgoing commitment to the moral importance of species and ecosystems requires more than just a few nice words about them. It will often demand that humans make sacrifices. This is particularly awkward for the ethicist, since moral progress up to this point has appeared synonymous with the increasing recognition and protection of human interests. We will require a very high standard of argument indeed before we seemingly turn back the clock of moral progress, placing the needs of cave-dwelling insects or ecosystems ahead of, or on a par with, children's educations or peasant farmers' food.

Nonanthropocentric and Anthropocentric Ethics

Philosophers have been working hard to meet this high standard of proof. The extant ethics of the environment tend to fall into two categories. Swiftly described, *anthropocentric* ethics trace the preservation of any chunk of nature back to some human benefit or interest, while *nonanthropocentric* ethics make the value of nature independent of any such benefit or interest.

A historically significant tenet of nonanthropocentric ethics is that natural collectives and individuals are valuable *intrinsically*. Intrinsically valuable things are usually held to be those things that have value regardless of any benefits they bring to other objects. Advocates of this interpretation often cite individual human beings as paradigms. Conclusive proof that a human is utterly incapable of bringing about some other desirable states of affairs would not void her or him of value. With this rough understanding of the concept, we can set up a contrast with the concept of *instrumental* value accorded a starring role in many anthropocentric ethics. Instrumentally valuable things are supposed to possess worth by virtue of their propensity to bring about some other valuable state of affairs. Money is a familiar example. Its value does not survive removal from a market economy.[4]

Making a pelagic ecosystem the moral equivalent of money seems more likely to lead to the views of overfishers and strip miners than those of environmentalists. However, the equivalence of anthropocentric value with crude instrumental value radically understates the sophistication of contemporary anthropocentrism about nature. For example, Bryan Norton (1987) makes a distinction between two types of anthropocentric value. He says that an object has *demand value* if it can satisfy some currently held preference. Those objects with *transformative value* don't necessarily satisfy any currently held preferences but instead give reasons to alter them. To take Norton's example, a teenager may choose to pay thirty dollars to see a rock concert. This means that the concert has a demand value of thirty dollars. She would not choose to pay anything to see a classical music concert, giving it a zero demand value. However, having attended the classical concert, she goes on to acquire a lifelong and emotionally enriching interest in Mozart and Beethoven. The classical concert, therefore, has high transformative value. Norton proposes that reasons for preserving, rather than using up, nature derive from its high transformative value.

Since I defend a nonanthropocentric ethic, it might seem my philosophical duty to pick holes in anthropocentric arguments. In fact, I will offer no extended discussion of anthropocentrism and, indeed, concede that there are good reasons for not gainsaying this moral strategy. We may need to call on anthropocentric arguments to sustain some parts of the environmentalist agenda that are not adequately supported by the nonanthropocentrism of this book.[5] This said, we should recognize a nonanthropocentric ethic, if achievable, as constituting the core of our moral approach to nature. The nonanthropocentric ethic defended in this book employs the same manner of argument we use to defend the worth of individual humans in defense of nature's worth. We think of such arguments as showing that an individual human has value independent of the needs and interests of other humans, placing this value ahead, in moral terms, of the many uses an individual may be put to by other human beings. So it will be with nature.

Conservative and Radical Roles for Science in an Environmental Ethic

Getting a nonanthropocentric environmental ethic off the ground will be no easy task. One problem stems from the necessarily interdisciplinary nature of the enterprise. It is widely recognized that any such ethic must pool the efforts of workers from many different fields. Action-guiding moral principles may be the preserve of philosophers, but we are not after any old ethic. The biological and physical sciences must come into play if any of our moral claims are to be relevant to the environment.

Since our roll call of contributors will include philosophers, ecologists, systematists, and evolutionary theorists, we need widely accepted principles to tell us how to transform disciplinary diversity into a single set of nature-respecting principles.

We now come to the second, more methodological goal of this book: an account of how information from science can lead to the revision of moral principles. The environment will serve as a testing ground for this revisionist strategy.

As things stand, environmental ethics have no uncontested principles to guide the integration of diverse findings and speculations. There are no agreed-upon views about how the discoveries and claims of one discipline can be joined to discoveries and claims from other disciplines to produce moral principles. Symptoms of interdisciplinary anarchy are apparent. There is deep controversy about what kind of enterprise environmental ethics is supposed to be. On the one side, philosophers pretend to find an environmental ethic simply by subjecting a grab bag of normative notions to armchair scrutiny, and on the other, scientists pretend to extract an ethic from ecological or evolutionary biological models alone.[6]

Compare the situation with another grand enterprise requiring cooperation from workers in diverse fields. There could have been no humans on the moon without rules telling us how to integrate the work of engineers, astrophysicists, and human physiologists. It is certain that the interdisciplinary anarchy found in the moral debate on the environment would have killed off the space program.

So how might moral theory and the sciences of the environment be combined to yield an environmental ethic? There are conservative and radical views about the methodological relationships between science and ethics. In the conservative view, moral philosophers are to discover fundamental moral principles that do not vary in the face of scientific information. Ecologists and evolutionary biologists can then assist in the formation of derived principles whose job is to make explicit how established fundamental ethical principles govern our interaction with nature. For example, having settled on "Persons ought to be respected" as a fundamental ethical principle, we might observe that the livelihoods of indigenous peoples are inextricably tied to native flora and fauna and so acquire a derived moral interest in this flora and fauna.

In the radical view, relevant science can help in the formation of new fundamental ethical principles. Precisely how this might happen, given the differences between descriptive scientific claims and moral claims, is a mystery. Any would-be radical should be alerted to the risks of tampering with the received relationships between facts and values. Making fundamental ethical principles sensitive to information from science threatens to magnify the interdisciplinary anarchy already bedeviling environmental ethics. Which science should we choose? How flexible should familiar ethical views be in the face of scientific information? Plainly, there will be many questions for the radical to answer.

My distinction between conservative and radical approaches differs from the more familiar distinction between anthropocentric and nonanthropocentric ethics. Whereas the traditional distinction divides ethics in terms of a substantive property—anthropocentric ethics tie all value back to humans, and nonanthropocentric ethics do not—the distinction between radical and conservative ethics is a purely methodological one. Radical ethics allow science a creative role; conservative ethics deny it such a role. This point made, anthropocentrists do tend to be conservatives. Some brand of anthropocentrism about value in general has been an often unargued premise since the dawn of what contemporary philosophers recognize as ethical theorizing, and conservatives deny that science can shake this assumption.

In spite of the anticipated dangers and difficulties, I take the second, radical option in showing how scientific discovery can transform an ethical orthodoxy that is, at best, indifferent to nature into an environmental ethic.

DEFINING INTRINSIC VALUE?

Of late, intrinsic value has acquired something of a fool's gold reputation among writers on ethics and the environment. Repeated failures to come up with a single widely accepted account has prompted a retreat from earlier ambitions to count nature's worth in terms of intrinsic value.[7] These difficulties stem from a particular way of viewing the defining task.

According to the *classical view of meaning,* a definition is a set of necessary and sufficient conditions for the correct application of a concept. In saying that the definition of "triangle" is "a figure completely bounded by three straight lines," the classical definer is holding that all and only triangles are figures completely bounded by three straight lines. Because most concepts are not so straightforward as "triangle," the classical view is usually paired with a tool for clarifying the necessary and sufficient conditions associated with more problematic concepts. This is *conceptual analysis.* We know that to assign intrinsic value to an object is to somehow give it salience in moral discussion, but what does this mean more precisely? The conceptual analyst begins with a candidate account of what it is for something to be intrinsically valuable. This is tested by generating imaginary stories that seek to separate the familiar notion from the suggested analysis. There are two ways in which thought experiments of this type can lead us to reject the proposed definition. The account may exclude objects that we are strongly disposed to adjudge intrinsically valuable, or it may include objects that we feel sure are not intrinsically valuable. Failure at this stage may prompt us to modify our analysis and resubmit it to trial by thought experiment—or to start from scratch.

Imagine that I were to claim that the necessary and sufficient condition for being intrinsically valuable is being human. The critic can bring two types of thought experiment to bear on this proposed definition. We might first imagine a being that is certainly not human yet is conscious, rational, forms projects, and cares about its own ongoing existence. Would such a being count as intrinsically valuable? Respondents whose intuitions have been trained by decades of *Star Trek* episodes will say yes. Perhaps then we could turn to something such as a one-day-old conceptus. It is certainly human, yet could such a thing, lacking any conception of itself or ability to suffer, count as intrinsically valuable? Many will answer no. Strong reactions to stories such as these stand in the way of any simplistic identification of the intrinsically valuable with the human and prompt the conceptual analyst to continue the search for an account.

Those with environmentalist sympathies hope to use conceptual analysis to unearth long-hidden connections between familiar moral concepts, on the one hand, and healthy ecosystems and thriving species, on the other. Environmental neglect will turn out to be abetted by a mistaken understanding or application of the notions of intrinsic value.

I reject the classical approach to the meaning of "intrinsic value." This is not because I reject conceptual analysis—later in this book I will make great use of this tool to determine how value is distributed throughout nature. My quarrel is, instead, with the classicist's interpretation of the results of conceptual analysis as necessary and sufficient conditions, together with the view that this semantic part of our inquiry is to be conducted in isolation from information about the natural world. The following chapters explore a more empirically oriented task for conceptual analysis, one that while targeted at the concept of intrinsic value differs from finding necessary and sufficient conditions for it.

There are two reasons for thinking that the classical approach to the notion of intrinsic value is unlikely to be of great help to the environmentalist. First, the efforts of skilled conceptual analysts notwithstanding, the search for necessary and sufficient conditions is short on results, especially in the contested moral domain.[8] Here's a quick glimpse of the contemporary controversy over intrinsic value.[9]

Earlier I suggested that intrinsic value stands in opposition to instrumental value. Assuming there are no other varieties of value, intrinsic value will then be noninstrumental value. Many environmental ethicists will be happy with this negative definition, hoping that to make a good case for intrinsic value is to rule out an exclusively instrumental treatment of nature. However, those following Christine Korsgaard (1996) reject any identification of intrinsic value with noninstrumental value. A thing's intrinsic good is good "in itself," and this for Korsgaard is correctly opposed not to instrumental value but to extrinsic value. She holds that an object is extrinsically valuable if it gets its value "from some other source."

A related but distinct account of intrinsic value comes from G. E. Moore who claims, "To say that a kind of value is 'intrinsic' means merely that the question whether a thing possesses it, and in what degree it possesses it, depends solely on the intrinsic nature of the thing in question."[10] For Moore, whether or not an object possesses intrinsic value is a matter of its intrinsic nature. The intrinsic nature of the object is, in turn, constituted by its nonrelational properties, properties that are possessed regardless of context.[11]

Other writers see the debate over intrinsic value as taking place at an altogether deeper level in ethical theory. For them, intrinsicalness is best contrasted with subjectivity. The subjectivist holds that human consciousness or attitudes are the sources of all value. Accordingly, the advocate of intrinsic value is interpreted as holding that there is some variety of value completely independent of human ethical appreciation.[12] Somewhat confusingly, the concept of intrinsic value appears sufficiently malleable to allow an overtly human-response-dependent reading. Eugene Hargrove (1992) challenges the orthodox pairing of intrinsic value with nonanthropocentrism. He hopes for an environmental ethic founded on an anthropocentric intrinsic value, in which this anthropocentrism is driven by a recognition of the response-dependent nature of all value.[13]

This is but a very brief excerpt of the larger debate on intrinsic value. Each widely defended account appears to have its own set of intuitive strengths and weaknesses.

Should we be worried about what might seem nothing more than a dispute among philosophers of language? I think so.

Moral terms fall into two categories. First, we have bedrock moral notions such as *justice, good,* and *right.* There are certainly no widely agreed upon definitions of such terms. An initially disinterested surveyor of the history of debate over *justice* will find that to act justly may be to act in accordance with God's will, to promote the best balance of happiness over suffering, or to follow rules that would have been agreed upon by contractors who knew nothing about their actual capacities and social positions. Though it is certainly the case that some analyses become passé as others become fashionable, the prolonged philosophical speculation on *justice* has led to an increase rather than a diminution of definitional diversity.

What are the practical implications of such disagreement? Centuries of philosophers' headaches notwithstanding, failure to agree on a precise meaning of *justice* has not threatened its tenure in moral discourse. The longevity of terms such as *justice, good,* and *right* ensures that definitional difficulties do not lead us to jettison them. It is one of the great strengths of anthropocentric environmental ethics that they are constructed out of such durable notions.

Failure to arrive at agreed-upon definitions is of greater concern for a notion in the second category, a notion petitioning for admission to the moral domain. I suggest that the conditions under which we would deny a term access are different from those under which a widely used term would be dislodged from an area of discourse. A degree of confusion about meaning that can be shrugged off by advocates of *justice* or *goodness* should prove fatal to arguments hinging on a novel notion.

Consider the parallel with scientific discourse. *Consciousness* and *life* are examples of vague-seeming descriptive notions with established folk-theoretic pedigrees. Though these terms may make some scientists uneasy, their widespread use means that they are not so easily ignored. Faced with them, the cognitive scientist or biologist will need to choose between equally awkward paths, either striving to show how they map onto entities with proven scientific bona fides or purging the relevant discourse of them. An altogether different view will be taken of terms that are both vague and relatively novel, such as *aura* and Rupert Sheldrake's *morphic resonance.*[14] Until precise links can be made with scientifically respectable notions, such language will not be worth bothering with.

The philosophical traditionalist may put the failure to arrive at consensus down to either insufficient imagination in generating accounts or the sloppy investigation of dreamed-up scenarios. However, the ongoing failure to come up with a definitive analysis seems to put the environmental ethicist's "intrinsic value" in the category of terminology to be avoided. It is a notion relatively new to thinking about nature. It has no agreed-upon links either to other moral terms or to the concepts of the environmental sciences.

But there is a second, more pressing problem for those who seek to harness the notion of intrinsic value to a revisionist ethic and whose preferred tool is conceptual analysis.

Simply put, conceptual analysis seems a more appropriate tool for those offering conservative ethics, ethics that do not seek to radically challenge our current intuitive ethical understanding. But a consequence is that even successful analyses of notions such as intrinsic value are unlikely to provide firm and extensive safeguards for the environment.

To bring this methodological conservatism into focus, we must look more closely at what we are doing when we analyze a commonsense notion. What drives our intuitive responses to thought experiments? According to a likely story, when we respond to imaginary stories about intrinsic value, we are being informed by our tacit moral theories, the theories that guide instinctual moral judging and behaving. In noting these responses, we hope to determine the nature of the conceptual entity or entities responsible for them and to make previously tacit principles explicit.[15]

The idea that the accounts arrived at by conceptual analysis are "in us" from the start poses a problem for the environmentalist. The apparent reluctance of most people to murder other people might indicate the existence of a tacit moral principle requiring some manner of respect for human life. This explains why theories that protect humans receive intuitive support. Contrast our intuitive reverence for human life with our attitude to a termite mound. Most people are unhesitatingly prepared to trade off high levels of environmental destruction for small and ephemeral improvements in human welfare.[16]

The risk, if we focus too much on the excavation of tacit moral theory, is that we will arrive at a fully systematized view of the intrinsic *valuelessness* of the environment. Such a procedure stands to bind value notions much more securely to the concept *human* than to the concepts *ecosystem* or *species*.

Earlier I said that I plan to find a task for conceptual analysis that differs from finding classical definitions. Detaching conceptual analysis from the classicist's search for necessary and sufficient conditions will not meet the above described worry. I concede that conceptual conservatism is the price of commitment to conceptual analysis.

My diagnosis of morality's human bias points to a strong connection between moral and psychological language. We think that having moral standing has a great deal to do with having some manner of mind. I won't challenge this connection. Rather, I will show how the abandonment of the meaning classicist's interpretation of conceptual analysis allows us to considerably extend morality's reach.

In Search of a Theoretical Definition of Intrinsic Value

Where do we find an alternative to the classical approach to intrinsic value? Science has done fine work in alerting us to the environmental dangers caused by human activities, and it also offers a path to environmental value by way of a *theoretical definition* of intrinsic value.

A feature of the classical approach to meanings is a sharp distinction between the stage of conceptual analysis, conducted from within the philosophical armchair, and the empirical stage of finding the objects covered by the given term, conducted from without it. The philosophical traditionalist wants a complete account of what it takes for something to be intrinsically valuable before setting out to say how this value is distributed throughout the environment. For the theoretical definer, by contrast, the defining task is continuous and interspersed with empirical investigation. We do some preliminary analysis of a notion. This tells us where to begin our search. We learn something about the natural world and then return to our initial analysis, perhaps modifying it, perhaps focusing on different definitional strands of the original notion.

Philosophers of language will have doubts about whether a theoretical definition can be the right tool for us. The notion seems too bound up with projects in metaphysics and the philosophy of science. There is a widespread consensus that theoretical definitions pick out *natural kinds*.[17] Later I will go into some detail on the notion of a natural kind, but for now I will say that a natural kind gathers together objects sharing some property of great importance in terms of scientific prediction and explanation. Possessing atomic number 79 or having been formed by natural selection for the purification of blood are such properties, as they are what interest chemists and biologists about gold and kidneys, respectively. The theoretical definition of *gold* will be "the element with atomic number 79," and of *kidney*, "the organ whose selected task is to purify blood." *Intrinsic value* seems quite unlike *gold* and *kidney* in this respect. It doesn't seem targeted toward natural kinds.

To produce a definitional tool appropriate for a moral term such as *intrinsic value*, it is necessary to weaken, in two ways, the connection between theoretical definitions and natural kinds.

First, and less importantly, a definition need not be right on the money about relevant kinds. At the dawn of our contact with a given range of entities, some definitional opacity will accompany our uncertainty about what our naturalistic inquiry is going to dredge up. We must start with something, and the theoretical definer may have only common sense or superstition to guide any initial attempts at understanding. What matters is a disposition to be sensitive to information about kinds, where it is available.

Imagine that someone is seeking to define gold prior to the emergence of analytical chemistry. A definition grounded in contemporary folk theory might include information about color, weight, malleability, potential to be created from lead by yet-to-be-discovered alchemic processes, and so on. These combined descriptions will serve as our definitional starting point, helping direct us toward relevant kinds. Though the modern way of defining gold as the element with atomic number 79 shares no commonality with that older account, it owes a great debt to it. Just as there could have been no modern humans without *Homo erectus,* so too our present natural kind–isolating definition of gold owes its existence to an older, more primitive theoretical definition that is vague and inaccurate in the light of today's science.

Even if the above comments are correct, we will have told only half the story about intrinsic value. Information about natural-kind boundaries could not possibly exhaust the meaning of *intrinsic value* as it might do for *gold* and *kidney*. While all have descriptive components, *intrinsic value* has a normative dimension entirely absent from *gold* and *kidney*. The normative component establishes appropriate connections to overtly moral notions such as "warranting respect" and "worthy of protection." In contrast, the descriptive component guides us in our application of the concept by pointing to things in the natural world, and it is here that information about kinds will be relevant.

If there is some systematic relationship between normative and descriptive components, then much of what goes for the refining of the concept *gold* may also go for the concept *intrinsic value*. Again, our inquiry begins with imprecise, widely shared folk understanding. This is then deepened with information from the sciences of the environment and ongoing moral theorizing. Though the normative component won't be mapped onto any moral natural kinds, progress with respect to the descriptive component can "drag" views about the normative component along with it. The result will be a notion clearly and precisely located among other moral and scientific notions.

There is an important way in which these definitional aspirations are less ambitious than those of the classical theorist. In deciding the meaning of the notion, the classical theorist hopes for an account that can be substituted wherever we find the words *intrinsic value*. I make no such claim. If we want an account of intrinsic value suitable for aesthetics, we may start with the same initial analysis. While the environmentalist finds semantic flesh in the sciences of the environment, the specialist in aesthetics will look instead to the study of art. An investigation of artworks will shape the descriptive component of our concept and, consequently—if the linkage thesis is correct—the normative component. Though notions of intrinsic value may have a common starting point, we can expect them to diverge rapidly. Not all areas will be best served by the same definition and modification of intuitive understanding of the basic folk notion, and a pluralism about intrinsic value seems the sensible option.

The fact that I am not only a philosopher interested in nature but also a philosophical naturalist will have a significant impact on the manner of moral advice I will offer. The defense of the basic moral worth of nature has some immediate practical implications. We will no longer be able think of the environment as a moral zero. Much of what humans do to nature is straightforwardly bad for living things, and the recognition of life value enables a straightforward and forthright response to toxic-gas-belching industries and human-caused deforestation, salinization, and desertification.

These blanket recommendations aside, I will be more cautious in practical than in theoretical matters. Naturalists are against philosophers' quick conceptual fixes. An inspection of our world-directed concepts may point us in the right direction, but it alone does not give us significant truths about the world. Scientific information inaccessible from the philosophical armchair is essential to the definition of core concepts and the formation of fundamental principles. So it should be with the moral appreciation of nature. The main substantive practical upshot of this book should be a reconceptualization of the sciences of nature. Finding out about the natural kinds constituting the environment not only leads to knowledge about how they are best fostered but also helps to show why they ought to be fostered. If a moral consideration of nature turns out to be continuous with the scientific understanding of it, armies of environmental scientists are unwittingly doing moral philosophy.

WHY ARE HUMANS MORALLY SPECIAL?

Our starting point is this: something is valuable intrinsically if it is valuable in itself, or regardless of its benefit to any other being. This account is not supposed to stand up to detailed scrutiny; it serves only as a springboard to a naturalistic story filled out with information about individuals and wholes making up the environment.

Many moral theorists have picked up on the intuition that humans, and humans alone, matter, regardless of the impact they have on other beings. Can any human property account for this moral salience? If we can't point to some sufficiently important natural difference between humans and nonhumans, we should not rest content with this exclusive moral focus on humans.

Some attempts to find a difference are unsatisfactory. For a long time people said that unlike keas, termites, and tundra ecosystems, humans were created in God's image and alone possessed immortal souls. In many quarters, an evolutionary account has displaced the theistic view about human origins. At the same time, many people have come to conceive of themselves not as beings infused with immortal, supernatural souls but rather as sophisticated physical objects standing alongside a host of other physical objects. Our suite of adaptations makes us more complex than other evolved objects in some ways, but less complex in others.

The debate between those who hold that science provides us with a picture of the world incompatible with God's existence and those who do not rages on. The mere fact that there is such a debate gives good pragmatic grounds for finding alternative support for human worth. Societies with no consensus about the existence or characteristics of a supernatural creator nevertheless require some moral consensus. For this reason, defenders of human moral dignity need to look outside religion.

The Psychological View About Intrinsic Value

One familiar way of assessing the moral difference between humans and the rest of nature is by way of the concept of *person*. Writers in moral philosophy have long held that there is something very important in being a person, and this notion has accordingly played a pivotal role in earlier moral revolutions. Individuals such as noncitizens and slaves, once lying beyond the boundaries of the moral community, came to be recognized as persons and were thus admitted to it.

The moral elevation of the concept of *person* does direct us away from humans to some extent. Just as *human* does not capture the necessary and sufficient conditions for intrinsic value, so too it fails to capture the necessary and sufficient conditions for personhood. Fictional Daleks and Vulcans seem to count as nonhuman persons, and anencephalic babies are good candidates for human nonpersons.[18]

Be this as it may, most humans are, or will become, persons, and the only uncontroversial persons hereabouts are humans. Perhaps we can set the environmental moral revolution off by urging that environmental individuals and systems are persons. Gary Snyder urges an "ultimate democracy" in which "plants and animals are also people, and . . . are given a place and a voice in the political discussions of humans. They are 'represented.' 'Power to all the people' must be the slogan."[19] Matters are not this simple. Plants and the vast majority of nonhuman animals are without a doubt not persons. This becomes clear once we ask what is central to our notion of personhood. What matters about persons is that we have minds. Certainly no plants and plausibly only a few nonhuman animals have minds.

To better understand its importance in moral theory, we will have to look more closely at our intuitive theory of the mind or *folk psychology*.

Folk psychology sounds a somewhat dismissive label—but it is simply a name for the set of principles, known more tacitly than explicitly, that guide us in our daily dealings with other persons. Folk psychology divides mental states into two categories. First, there are intentional states, states whose distinguishing mark is meaning or content. Most familiar among these are beliefs and desires. Contents are at the heart of folk psychology's story about behavior. A desire that very reliably causes the ingestion of gazpacho is one with a very special content; there is a good chance it is the desire to eat gazpacho. The belief with the content that the Plaza del Sol is not far from Gran Via is one that equips its possessor most reliably in a world in which Madrid's main road, Gran Via, is near its central point, Plaza del Sol. On the other side of a famous philosophical divide come qualitative states. Red afterimages and pains are distinguished by their qualitative aspect or "feel." Many think that qualitative states are irreducibly subjective, taking them to derail physicalism about the mind.[20]

Both varieties of mental state are invoked in attempts to separate the morally worthy from the morally worthless. We can identify two broad lines for how psychological states make something morally important: the *rationalistic* and the *hedonistic* views. The rationalist selects from the intentional side of the commonsense divide. Contents are essential to the plans, projects, and desires that a person uses to capture the moral good of a human being. We are important because we can *plan* to become corporate lawyers or *hope* to appear in Hollywood movies. Contents are also fundamental to the beliefs that guide us in achieving our goods.

The hedonist focuses on qualitative states, saying that it is the human capacity to experience pleasure or pain that matters about us. In a well-known statement of the view, Mill says, "Pleasure and freedom from pain, are the only things desirable as ends."[21] Pinching a human can cause pain, foot massages can cause pleasure, and according to the hedonist, it is the job of morality to take account of such experiences.

Everyday thinking about the moral importance of humans combines elements of both rationalistic and hedonistic reasoning, and I am not interested in arguments that seek to decide between them. I will use the name *psychological view* to pick out the range of theories, rationalistic and hedonistic, that use some collection of folk psychological notions in sketching the boundaries of intrinsic value. It is this second-order view that will be my main focus.

If sustainable, the psychological view provides firm grounds for human moral superiority over most of nature. Plankton may be essential to almost all aquatic life, archaeopteryx may be an evolutionary watershed, but we are rational, feeling beings and are thereby nearly unique in our intrinsic value. A kauri's lack of a complex central nervous system makes it very unlikely that cutting it down causes it pain, and kauri trees have no desires or projects that the lumberjack's ax can frustrate.

Environmentalists have been quick to challenge the psychological view. Indeed, we do run into a problem in stipulating a strong connection between descriptive theory about what kinds of things have minds and theory about moral value. As will become clear in the next chapter, certain widely discussed arguments about the relationship between descriptive and normative notions make it unlikely that we will find any logically coercive proof of a link between the moral and the folk psychological. Logic alone cannot mandate the move from "Humans are more rational than rabbits" to "Humans are more morally important than rabbits." However, even if it is not a matter of logic, the connection between the moral and the folk psychological is very deeply ingrained in us. Folk psychological notions have cohabited with moral notions for so long that they have come to appear quasinormative. The interconnections are so dense and complex that it is next to impossible to imagine what it might be like to rebuild morality purged of any psychological vestige. Our initial account of intrinsic value needs to acknowledge the entrenched role of folk psychological notions.

To summarize, the psychological view of intrinsic value holds that for something to be valuable in itself, regardless of its benefit to any other being, it must possess some set of folk psychological states. The problem for the environment is that the nearly twelve billion feet busy trampling the life out of ecosystems belong to the most uncontroversial deservers of moral value. In the next chapter, animal welfarists raise issues about the possession of minds by some nonhumans. If we accept their arguments, we may embrace apes, pigs, and even chickens as having incontrovertible intrinsic value. The bad news is that the kinds of things most of interest to environmentalists seem unsalvageably nonpsychological. Does this leave them beyond morality's reach?

CHAPTER 2

Science's Bridge from Nature to Value

SCIENTIFIC FACTS AND VALUES

The chief asset of anthropocentric ethics is that they require no challenge to the moral primacy of consciousness, language use, rationality, or other mental traits. This means that anthropocentric reasons are readily recognizable as moral reasons. It seems that nonanthropocentrists have no option but to challenge the psychological view about intrinsic value. Here the going gets tough.

I have described our familiar concept of intrinsic value as having two components, descriptive and normative. The normative component locates intrinsic value among other moral notions, telling us that intrinsically valuable things are valuable without reference to other things. According to our preliminary take on the descriptive component, certain types of folk psychological state are required for this kind of value.

This account of the anatomy of intrinsic value immediately prompts some questions. What is the nature of the glue binding together these two disparate-seeming components? Later in this chapter I will run through David Hume's (1978) arguments against there being any logical link between the descriptive and the normative. If such arguments are successful, we might conclude that the currently fashionable bond between psychology and morality is logically arbitrary. We will allow that had viruses been making up ethics, they would have been equally justified in making speed of replication morally salient.[1]

The conclusion that the connections between the moral and the folk psychological at the heart of moral common sense come down to nothing more than brute associative habits would not be all good news for the moral radical with environmentalist sympathies. Problems in justifying the currently favored pattern of normative-descriptive linkages will carry over to any novel pattern of linkages. If no pattern turns out to be rationally defensible, then some form of intrinsic value eliminativism might be the best option. However, supposing that we do feel the need for a scheme for mapping the normative onto the descriptive, then there would seem no good reason not to stick with the tried—albeit not proven—mapping at the heart of moral common sense. Lacking the argumentative resources to budge us from some version of the psychological view, brainwashing might turn out to be the only tool at the disposal of the would-be reviser of intrinsic value.

I will argue that science can take us from a starting point described by the psychological view to an environmental ethic, thereby bridging the gap between familiar value concepts and nature.

Many people who write about the environment will resist proposals to give science any significant role in ethical argument. Opponents of science fall into two camps. Some think that the gap between science and morality is bridgeable but see only bad outcomes. Whether by rational or irrational means, we will end up translating scientific and technological imperatives into moral principles, and these principles will inevitably be hostile to nature. Others think that scientific information is simply not of the right type to bring about fundamental moral change. Arguments for value change in response to science fail to be rationally coercive—meaning that our values can and ought to remain invariant in the face of any manner of discovery about the world. Those convinced by the arguments of Hume and G. E. Moore (1922) fall into this category.

Science as Moral Peril

I begin with a quick examination of the views of those who think science gives rise to moral imperatives opposed to nature. A familiar explanation for this hostility is that science fails in its aspirations to value neutrality and objectivity; it is secretly normative. According to some, the scientific method cooks up its own moral "oughts." As atomistic science cuts up nature in order to understand it, science cannot help encouraging us to cut up and recombine it. Others emphasize science's propensity to absorb the values of the society that makes it. Biologist Richard Lewontin fingers science as the "chief legitimating force in modern society."[2] He says science "reflect[s] and reinforc[es] the dominant values and views of society at each historical epoch."[3] A favorite example of writers with similar views is the relationship between early Darwinism and Victorian colonial and class consciousness. Darwin's theory, with its "survival of the fittest" talk, seemed to vindicate popular prejudice about the underclass and non-Europeans. Environmentalists will be particularly concerned about the impact of the values of late-twentieth-century capitalism on science. If science is nothing more than a fig leaf for industrialized capitalism, nature's chief despoiler, what hope do we have that it could yield real solutions to environmental problems?

One doesn't need to believe that science is secretly normative to think that it represents a danger to morality. Physicist Wolfgang Pauli and, more recently, journalist Bryan Appleyard worry not so much about science's propensity to lend credence to the morally unworthy but instead about its across-the-board delegitimizing influence. Science's inherent valuelessness rather than its value-ladenness is now the problem. Pauli forecasts apocalyptic consequences as science sweeps away the bases of moral belief:

> At the dawn of religion all the knowledge of a particular community fitted into a spiritual framework, based largely on religious values and ideas. The spiritual framework itself had to be within the grasp of the simplest member of the community, even if its parables and images conveyed no more than the vaguest hint as to their underlying values and ideas . . . That is why society is in such danger whenever fresh knowledge threatens to explode the old spiritual forms . . . In western culture, for instance, we may well reach a point in the not too distant future where the parables and images of the old religions will have lost their persuasive force even for the average person; when that happens, I am afraid that all the old ethics will collapse like a house of cards and that unimaginable horrors will be perpetrated.[4]

Bryan Appleyard (1992) shares Pauli's concerns. His is a book-length account of science's corrosive effect on morality. Appleyard describes the conflict between the culture of science and traditional culture with the help of a map metaphor. The maps of nonscientific or traditional cultures

> show some regions with a reasonable degree of certainty. But then knowledge fails and the imagination of the mapmaker takes over. The region of the known shades away into myths and fairy stories—dragons and giants at the world's end, a landscape of chaos beyond the limits of

order. There was a line drawn to mark the limits of human knowledge. There was an outside, a beyond.[5]

We err if we see the "beyond" as mere space to be filled in. Appleyard's dragons and giants, or Gods and spirits, are the bases of morality. He says that the maps of liberal-scientific culture are "complete and clear; there is nothing missing and there is nothing we cannot understand"; consequently, they find no place for the traditional bases of ethical belief.[6]

Such alarmist conclusions seem unsustainable. However we end up resolving the philosophical riddles surrounding nature's good, close contact with the "beyondless" maps of science seems not to have prevented scientists from seeking this good.[7]

Appleyard does allow that, at least with respect to nature, some moral benefit can be derived from science. The trick is to be careful about what science we lean on. A cursory examination of recent history does turn up a great deal of science and technology that has been bad for nature. The Hiroshima bomb and leaky Soviet nuclear submarines could not have been built without early-twentieth-century physics, and it is no surprise that examples such as these feature prominently in doomsaying speculation. Appleyard sees a holistic ecology as the antidote to nuclear physics. Ecology is "a way of turning science against itself,"[8] dampening the industrializing imperative of liberal-scientific humanity by exposing humankind's technological urges as suicidal.

Appleyard's is a good news–bad news story, however. The same methods that he thinks demonstrate nature's anthropocentric value render it nonanthropocentrically valueless. Ecology is incapable of locating intrinsic value in nature: "The environmentalist may enthuse about the peace of mind he may attain through correctly green behaviour. But, at base, his reasons for that behaviour are purely practical. There is no transcendent rationale. . . . We can only undo the harm we have done; we can aspire to nothing higher."[9]

In this book I aim to refute Appleyard's pessimism. Science helps show that nature is valuable in itself.

The Gap Between Facts and Values

Having briefly canvassed the arguments supporting the global moral nastiness of science, I return to the question of whether there could be a significant role for science in the hunt for environmental intrinsic value.

We are now in a position to identify two ways in which science might provide us with a concept of intrinsic value more generous to the environment. What I will call the ambitious approach focuses immediately on the relationship between the normative and descriptive concepts of our familiar concept of intrinsic value. Scientific information will be applied directly to the pattern of normative-descriptive linkages at the heart of the psychological view, with an eye to bringing about a new eco-friendly pattern.

The most obvious way for this to happen will be if some variety of ethical naturalism turns out to be true. Ethical naturalism is the view that normative properties are really natural properties and that the supposedly dual nature of our customary concept of intrinsic value is nothing more than illusion.[10] Past ethical naturalists have sought to identify moral properties with natural properties ranging from happiness to evolutionary success. If moral facts are some kind of evolutionary fact, then it is clear how the relevant science would lead to a revision of morality. The pattern of normative–folk psychological relations favored by common sense will be replaced with a set of normative–evolutionary linkages. Depending on the type of evolutionary ethic opted for, rather than saying that folk psychological things are valuable in themselves, we might instead say that things favored by natural selection are valuable in themselves. Later in this chapter I will have more to say about evolutionary ethics.

The second way of giving science a significant role in moral revision seems to offer more modest rewards. I will call this the cautious approach. Because the exclusive target here is the descriptive component of our concept of intrinsic value, rather than normative and descriptive components considered together, no manner of ethical naturalism is required. We do not question the value-anchoring role of folk psychological notions but instead scrutinize the part of psychological common sense that tells us what types of things have minds. According to this proposal, science redraws the boundaries of the category of intrinsic value by offering us a new and better psychological theory. A given range of objects may now be widely judged not to possess the required value-endowing psychological states and, therefore, not figure directly in a community's moral calculations. As science overturns views about the mindlessness of these objects, it shows them to be morally valuable.

I start by examining the cautious approach to ethical revision, and only when this seems incapable of supporting an adequate environmental ethic will I turn to the more risky ambitious approach.

THE LIMITS OF ETHICAL EXTENSIONISM

The most pressing initial need for the cautious revisionist is some story about what is going on when science exposes as mistaken a theory about a psychological category. To tell this story I must take a detour through some contemporary metaphysical theory.

I am going to treat psychological states as falling into *natural kinds*. As I suggested in chapter 1, a natural kind groups together objects with an *objective sameness*. Gold atoms make up a natural kind in virtue of the fact that a gold atom here has the same number of protons, electrons, and neutrons as a gold atom anywhere else in the universe. This microstructural property differentiates gold atoms from any superficially similar silver or iron atoms, thereby forming the foundation for explanation, prediction, and laws.

What moral payoff ought we to expect from this bit of metaphysics? The attention to appropriate natural kinds is supposed to ensure that our moral concern is consistent. We ought to reject a historically cherished view if it places in the same psychological category states that are not kind-mates or fails to group kind-mates together. Changes will matter morally when any folk errors we uncover exclude from moral consideration beings genuinely capable of suffering or planning to do things or include the non-sentient or nonintentional and nonplanning.

There are, however, reasons to doubt that the notion of a natural kind can really perform the moral-theoretic task of ensuring that our concern is consistent.

Up till now physics and chemistry have been salient in philosophical discussions about kinds. These are the sciences that philosophers hope will be, in an important sense, fundamental, capturing the structure of the world at its deepest level. When a given microstructural property is the hall-mark of a kind, there really is the promise of exceptionless laws. Yet psychological, rather than physical or chemical, kinds will be the focus of the cautious revisionist, and things become more muddled in the higher-level or "special" sciences. Some philosophers doubt that psychological and biological names pick out any manner of natural kinds. For example, Jerry Fodor (1968) thinks that the physical heterogeneity of states belonging to a psychological category distinguishes them from any natural kind. Other writers favor the wider usage of the term that I will rely on.[11] The notion of natural kind certainly has utility in special sciences such as biology and psychology. What matters is that the members of the category proposed for kindhood are similar enough in relevant ways for significant generalizations true of one member to be true of any other. We will expect less fundamental, special sciences to have less stringent overall requirements for objective similarity; biological functional kinds will often be quite chemically and physically heterogeneous. For example, gizzard stones make up a biofunctional kind or category despite the facts that different gizzard stones belong to different physical kinds and that some gizzard stones may be identical in physical respects to non–gizzard stones. Consequently, laws—if they can even be deemed such by linguists—will be riddled with exceptions. Despite such physico-chemical dissimilarities, gizzard stones all fill the same role within an organism's internal economy. When swallowed by a bird, they have the purpose of aiding in the digestion of food.

We need to think of a psychological kind as a grouping together of states with a sameness relevant to psychological explanation. Evidence for kinds roughly coincident with folk psychology comes from the theory's history of successful prediction and explanation. Either this success is a fluke or it rests on some variety of objective similarity between states with similar folk contents. So long as we can accept that common sense settles on genuine objective samenesses rather than just projecting them onto the brain, and that these commonalities form the basis of accurate prediction and explanation, we have a kind. It won't matter that psychological samenesses don't go very deep and are, perhaps, only capturable at some higher, functional level.

Though this tentative vindication of the categories of psychological common sense certainly depends on a rather liberal account of natural kinds, it does leave some room for dissent. If eliminative materialists are right, everyday psychological language fixes on no objective samenesses whatsoever.[12] Objective support for everyday prediction and explanation is an illusion.

The truth of eliminativism about folk psychology must have implications for moral theory. Traditional morality uses commonsense psychological notions to draw the boundary between the morally considerable and the inconsiderable. Once we banish intentional notions, this way of making the distinction will, presumably, have been shown to be unprincipled. If we are not special because we believe, desire, make plans, and so on, then what makes us special? Perhaps we will discover that any attempt to erect a boundary between humankind and the rest of nature will violate the requirement that we treat kindmates similarly. Will the category of the morally considerable be enlarged or diminished? Does folk morality default to almost everything, or does nothing turn out to be valuable? Keith Campbell (1986), for one, thinks morality may share the fate of folk psychology. He doubts that any displacing theory will have the resources to underwrite the moral enterprise.

Many writers have rallied to folk psychology's cause.[13] Though it would be inappropriate for me to review argument and counterargument here, I do think that a more probable outcome is that the theory that has guided us well enough for millennia is not wrong root and branch. In all likelihood, some pattern of objective psychological samenesses will be close enough to the folk pattern for intentional notions to be usefully retained through a rather more gradual, scientifically guided revision of the surrounding theory. The psychological view about intrinsic value requires that the boundaries of the moral community track this revision.

Some writers have already explored one possible revisionist outcome. Rather than attaching to no natural kinds, many folk psychological notions may be tied to kinds that are much broader than previously thought. Certain widespread practices such as factory farming and animal experimentation reveal that we consider animals to be of negligible intrinsic value. Commonsense psychology, at least initially, seems to support this. We feel some awkwardness in applying notions like belief and consciousness to languageless nonhumans.

This awkwardness translates into resistance to the idea of psychological kinds straddling species barriers. However, a blanket ban on animal minds is certain to be a mistake.[14] Bovine brains contain the same type of neural architecture that we strongly suspect to underlie our experiences. Establishing that animal and human heads are in important respects alike is a first step toward showing that animals have experiences similar enough to our own to fall into the same psychological categories. There is some room for doubt. Thomas Nagel's (1974) famous argument against materialism hinges on the likely strangeness of the perceptual experience of echolocating bats. The same could well go for bat suffering. Putting to one side Nagel's larger goal of disproving materialism about the mind, does our perspective toward bat, chicken, and rat pains prevent us from classifying them with our own pain? Not if bat pain is so unpleasant for the bat that it is strongly motivated to avoid it. According to Marian Stamp Dawkins (1993),

> animals may be completely different from us in their likes, and dislikes . . . But if we assume that they are like us in just one important

respect (namely that they too have unpleasant subjective experiences when they are prevented from doing things that they are strongly motivated to do), we have a bridge between our subjective world and theirs.[15]

Suffering and its expectation have similar implications for behavior in humans, bats, and chickens. Just as humans try hard to avoid torture, hens go to great lengths to avoid the wire mesh floorings on which they are forced to lay. This and other similarities support our use of the term "suffering" to describe the psychological causes of both behaviors. If Dawkins is right, it follows that a moral distinction between human and animal pain is not kind-respecting.[16] Such arguments have even more bite when they apply to apes, creatures whose brains plausibly contain kindmates for a wider range of value-endowing psychological states.[17]

Animal welfarists propose expanding the category of the intrinsically valuable so as to better accord with psychological natural kinds. This natural-kind approach may also lead to some shrinkage in the currently accepted category of the morally valuable. Consider the debate over the moral status of the fetus. Abortion liberals such as Michael Tooley (1972) and Peter Singer (1993) urge that an assessment of the cognitive capabilities of the fetus is relevant to the moral treatment of it. They say that since the fetus cannot conceive of itself as a being with a past and a future, we do nothing wrong in painlessly killing it.

We have seen how commitment to the psychological view and sensitivity to appropriate scientific information can lead us to revise our ideas about what is, and what is not, intrinsically valuable. In the next chapter I will focus on some problems for the bonding of value-endowing notions to specific kinds. For now I pass over these difficulties. The question we must ask first is what this approach offers to the apparently nonsentient organisms and biotic wholes that take pride of place in an environmental ethic?

Environmentalists certainly often talk as if the things they care about have minds. R. Edward Grumbine (1992) describes the transformative effect an expedition to the Greater North Cascades in Washington State has on students in an ecosystem-management course. He says that spending time in the Cascades and "participating in the local natural rituals" teaches us to "think like a mountain":

> To think like a mountain is to perceive different levels of biodiversity. The biosphere is present in the horizon's curve and in the banks of atmos-

pheric clouds. . . . The track of a solitary grizzly bear crossing a snowfield brings news of species and populations. . . . Koma Kulshan embraces all these levels of biodiversity at different scales of space and time. The time of hatching mayflies is meshed with the time of melting glaciers.[18]

Probably Grumbine does not seriously mean to ascribe a mind to a mountain. However, others have enthusiastically endorsed the idea of an intentional, phenomenological nature. Lynn White Jr., among others, suggests that before Christianity, animism was widespread. According to this animism, "every tree, every spring, every stream, every hill had its own genius loci, its guardian spirit."[19]

This view about the distribution of psychological states had straightforward ethical consequences:

> Before one cut a tree, mined a mountain, or dammed a brook, it was important to placate the spirit in charge of that particular situation and keep it placated. . . . By destroying pagan animism, Christianity made it possible to exploit nature in a mood of indifference to the feelings of natural objects. . . . The spirits *in* natural objects which formerly had protected nature from man, evaporated. Man's effective monopoly on spirit in this world was confirmed, and the old inhibitions to the exploitation of nature crumbled.[20]

It will be clear how the animists' psychological theory differs from what is currently widely accepted. Animists hold that rivers and mountains have folk psychological states while we deny that they do. White is correct in thinking that animism could have provided moral protection for nature. The bad news is that this view is not an option for us. A set of moral requirements grounded in animism is wrong because the psychological theory on which it is grounded is mistaken. There are two diagnoses that might be made. Perhaps pre-Christian common sense holds that psychological states are properties of immaterial soul or mind stuff, and that this adheres to oak trees just as surely as it does to human bodies. In fact, there is no such stuff. Perhaps the error takes a physicalistic form. Pre-Christian common sense may assert that the physical contents of a dandelion are as worthy deservers of psychological names as the physical contents of human skins. This is mistaken because human and dandelion innards are too different in terms of their actual and potential causal effects for the psychological theory that describes us to be true of a dandelion.

The failure of animists' attempts to find intrinsic value in nature appears to have grave implications. The naturalistic investigation of the environment finds nothing to play the same roles in kumaras or rivers that phenomenological and intentional states play in us. E. O. Wilson estimates that 93 percent of the animal mass in an average hectare of dry land consists of invertebrates.[21] Though we may discover that terms like *pain* and *enjoyment* describe some aspects of canine experience, the billions of environmentally crucial invertebrates are extremely unlikely to be sentient. Worse luck for them. Singer has no doubts about the consequences of this argument for the environment: "To extend an ethic in a plausible way beyond sentient beings is a difficult task. . . . Sentient beings have wants and desires. . . . There is *nothing* that corresponds to what it is like to be a tree dying because its roots have been flooded."[22] The absence of experience grounds Singer's contention that the nonsentient cannot be harmed or benefited and should, therefore, only figure in moral theory insofar as what happens to them affects sentient beings.

The nonsentience and nonrationality of most of nature is one problem. Even if we did discover that sufficiently many environmentally important organisms were sentient or rational, we might still be some distance from an adequate environmental ethic. What many people hold as distinguishing an environmental ethic from more familiar ethics is a concern with *wholes* or *collections* of certain types—species, ecosystems, the biosphere. Animal welfarists are avowedly individualistic. Singer is keen to dislodge speciesism, the view that facts about species membership make a difference to the moral worth of an individual. These facts have nothing to do with its capacity for suffering. Yet for the environmentalist, whether or not an individual is a member of the last breeding pair of a species or is an introduced pest is of central importance. J. Baird Callicott articulates this view, at the same time pointing out the limitations of any morality founded in psychology:

> The contemporary animal liberation/rights, and reverence-for-life/life-principle ethics . . . [provide] no possibility whatever for the moral consideration of wholes—of threatened *populations* of animals and plants, or of endemic, rare, or endangered *species,* or of biotic *communities,* or most expansively, of the *biosphere* in its totality—since wholes per se have no psychological experience of any kind.[23]

The moral revolution sought by animal welfarists demands careful attention to the boundaries of natural kind, and it is precisely this emphasis that seems to stand in the way of moral progress on the environment. We will infer the valuelessness of nature from its mindlessness. Though an emphasis on psychological kinds may bring us closer to chimps, pigs, and even chickens, it takes us further away from blue-green algae and Arctic ecosystems.

BEYOND ETHICAL EXTENSIONISM?

Ethical extensionists seek to reshape the long-accepted boundaries of our familiar concept of intrinsic value by modifying its descriptive component. Though they may manage to show that some previously disdained objects are intrinsically morally valuable, they do not manage to underwrite concern for the environment. What's needed is a more ambitious proposal for greater moral inclusiveness. As I explained in the opening section of this chapter, these proposals should be understood as using science to directly tackle the linkages between the normative and the descriptive components of our commonsense concept of intrinsic value. Science is called upon to cook up a new, more environmentally friendly set of descriptive-normative connections.

Earlier, I suggested that the ambitious revisionary strategy requires some version of ethical naturalism. Ethical naturalists of all varieties must confront the arguments of David Hume and G. E. Moore.[24] Hume contended that it is not possible for a valid argument to move from purely factual premises to a normative conclusion. We cannot derive the conclusion that deforestation is wrong from the claim that if deforestation continues, biodiversity will be reduced. What is missing is the separate normative claim that it is wrong to reduce biodiversity. Moore's closely related argument lays the charge of the naturalistic fallacy at the feet of would-be identifiers of good with natural properties. Suppose we take an interest in a given ecosystemic property P and we discover that a given natural system exhibits P. This does not settle the issue of the value of the natural system, as it can still be meaningfully asked, "Is it morally good for the ecosystem to have P?" Moore would have said that this shows there is more to being morally desirable than just possessing P. Moral concepts are inherently action-guiding; to recognize that something is morally valuable is to be given some reason to promote or protect it. Saying that something is worth preserving or promoting, therefore, goes far beyond locating some purely natural property in it.

The arguments of Hume and Moore have been very influential in Western thinking about morality. They spell bad news for our ambitious revisers of the concept of intrinsic value, as anyone who is struck by the gap between values and facts will likely perceive a gulf between values and those facts accessible to science.[25] This is well illustrated by the different ethical conclusions that have been drawn from claims about the precarious state of many natural systems. For example, experts tell us that overfishing is reducing marine stocks. Fishing grounds in the Atlantic, Pacific, and Mediterranean have seen declines of up to 60 percent on peak catches.[26] Environmentalists use facts like these to back up calls for a reduction in the numbers of fishing boats, together with restrictions on some of the most undiscriminating fishing technologies. Charles Hall (1990), on the other hand, describes a response by an economist that differs quite markedly from that of environmentalists. The economist uses considerations about economic efficiency to support demands that fish be harvested until the resource is entirely depleted. When there are no more fish, the money behind the fishing industry can be withdrawn and reinvested in other areas, thereby maximizing the economic potential of the capital.

Advocates of both conclusions agree about the relevant environmental science. The moral difference between the responses of the environmentalist and the neoclassical economist cannot be traced back to any disagreement about the way the world is. How, then, are we to derive an environmental ethic from what ecologists or any other environmental scientists tell us?

In what follows I will not attempt any detailed analysis of the barrier between facts and values. Rather, I will treat it as a hurdle in the way of attempts to get an ethic of the environment out of science. The points raised in this section constitute a strong prima facie case against science as a reviser of ethical belief, placing the onus of proof on those who see a connection between the concepts of the environmental sciences and normative concepts.

Is the Love of Nature in Our Genes?

Perhaps human nature is the best place to look for facts to ground the valuation of wild nature. Wilson (1984) claims that we have an evolved love of nature, or *biophilia*. This is a selected propensity to love the environment that has nurtured our evolution, and it is supposed to ground our recognition of the intrinsic value of nature.

Before outlining Wilson's argument, I will distinguish it from those appeals to evolution embedded in a more conservative, firmly anthropocentric line of thinking.

We know that natural selection has designed organisms for specific environments and that the systems that make up their bodies function best when they are in these natural habitats. Remove a crab from its shoreline environment, place it in the desert or snow, and things will very likely not go well for it. Equally, in destroying nature, we create surroundings for which we ourselves are ill adapted. Appreciating the facts about the evolved capacities of our bodies should lead us to expect, sooner or later, the same fate as the crab newly landed at the South Pole.

Wilson thinks our bond with nature goes far beyond the intellectual recognition of dependence. Humans have emotional, aesthetic, and even spiritual cravings to be close to and to preserve nature. These support our assignment of intrinsic value to nature. Wilson urges investigation of this "innate tendency to focus on life and life processes": "Humanity is part of nature, a species that evolved among other species. The more closely we identify ourselves with the rest of life, the more quickly we will be able to discover the sources of human sensibility and acquire the knowledge on which an enduring ethic, a sense of preferred direction, can be built."[27] Doesn't current indifference to the environment show that the road from hunter-gatherer to city-dweller has erased any bond with nature? Not so, says Wilson: "When humans remove themselves from the natural environment, the biophilic learning rules are not replaced by modern versions equally well adapted to artefacts. Instead, they persist from generation to generation, atrophied and fitfully manifested in the artificial new environments into which technology has catapulted humanity."[28]

Wilson claims to find evidence in modern minds for the persistence of biophilia. Ironically, the first piece of evidence for an innate love of nature is a specific pattern of natural phobias. These demonstrate the selective design of human psychology that enable us to deal with threats appropriate to a natural setting.

> People acquire phobias, abrupt and intractable aversions, to the objects and circumstances that threaten humanity in natural environments: heights, closed spaces, open spaces, running water, wolves, spiders, snakes. They rarely form phobias to the recently invented contrivances that are far more dangerous, such as guns, knives, automobiles, and electric sockets.[29]

Positive biophilic urges even manage to play some role in directing us within our urban environments. City-dwellers tend to be drawn to areas that best simulate natural settings: "The favored living place of most peoples is a prominence near water from which parkland can be viewed. On such heights are found the abodes of the powerful and rich, tombs of the great, temples, parliaments, and monuments commemorating tribal glory."[30] Natural areas are favored as holiday and recreation destinations: "A large portion of the populace backpacks, hunts, fishes, birdwatches and gardens. In the United States and Canada more people visit zoos and aquariums than attend all professional athletic events combined."[31]

Two questions arise. First, how likely are psychological adaptations such as those described by Wilson to exist? Second, how would any description of them support an intrinsic value–attributing environmental ethic?

Biophilia is one of a cluster of proposed Darwinian psychological adaptations. Sociobiologists and Darwinian psychologists have sought to explain traits and propensities ranging from the three-dimensional structure of color space to mate choice and distrust of those who look different from ourselves.[32] The epoch in human prehistory of particular interest is the Pleistocene, running from about two million years ago to a few thousand years ago. The African environment of the time is labeled the Environment of Evolutionary Adaptedness. According to Darwinian psychologists, this extended hunter-gatherer period allowed the probably gradual natural selective design of complex psychological structures. In proposing a Darwinian psychological adaptation, we want to show that it could have consistently met the challenges posed by the Environment of Evolutionary Adaptedness.

Biophilic responses can only have evolved by natural selection if they boosted the inclusive fitness of those who expressed them.[33] This means that in the Environment of Evolutionary Adaptedness, if not now, individuals with innate tendencies to love nature would have given their genes better chances of being represented in succeeding generations than non-biophilic competitors. Why should this occur? Wilson would have us believe that those blessed with biophilia better understand the threats posed by and opportunities offered by nature.

Selectional stories about the eye or heart begin with a relatively clear understanding that the mechanisms exist in the first place, of roughly what their physical structure is, and of how they interact with other systems. It could be psychological adaptations pose difficulties in each of these areas. Are the psychological traits Wilson is talking about robust in the way that we might expect adaptations to be, or are they more likely to be ephemeral patterns of thought provoked by novel aspects of our urban environments?[34]

Let's assume for the moment that Wilson can make a case for biophilia as a psychological adaptation. There are two further hurdles for him to jump. First, Wilson must show that a theory about psychological mechanisms has ethical implications. He must next demonstrate that the ethic justified by reference to human biophilia is one that affords the right manner of protection to nature, that it really is an environmental ethic.

How does Darwinian psychology stand up as a scientific unearther of ethical truths?[35] Herbert Spencer's evolutionary ethic provides Moore with one of the most flagrant committers of the naturalistic fallacy. Those competent in the use of moral language certainly have no difficulty in separating the idea of good from the idea of evolutionary success. More recently, interest in evolutionary explanations of psychological traits has prompted renewed interest in evolutionary ethics. Among the most forceful recent advocates has been Michael Ruse (1995), a sometime coauthor of Wilson's. The following discussion of Ruse's argument will help us to understand the moral force that biophilia could derive from its evolutionary roots.

Ruse is keen to distinguish his view from the more outrageous ethical claims extracted from evolution. Some evolutionary ethicists propose a range of novel and strange normative principles. One thinks of purported evolutionary justifications for large social inequalities and various nasty twentieth-century nationalisms and, more recently, Wilson's duty toward the global gene pool.[36] This is not Ruse's game. He thinks that an evolutionary understanding will leave our normative ethics largely untouched. Ruse says, "The job of the moral philosopher is not to prescribe some new morality, but to explain and justify the nature of morality as we know it."[37] Evolutionary theory is supposed to explain the pattern of norms that we accept as obvious. Does this make Ruse's evolutionary ethic an altogether unsuitable foundation for a revisionist environmental ethic? Not necessarily. Though Wilson's nature-respecting morality may seem entirely unfamiliar to contemporary ethical debate, he claims that the roots of this morality have always existed deep within us. Biophilia is supposed to be an ineliminable part of our psychological makeup, and Wilson aims only to make explicit the innate moral principles that traditional ethics have ignored.

Ruse thinks that evolutionary biologists hold the key to meta-ethical riddles bound up with moral objectivity. He paints this objectivity as nothing more than gene-imposed illusion. This realization could have unwelcome consequences. Ethics is "an adaptation, put in place by our genes as selected in the struggle for life."[38] He asserts: "Ethics is without justification or foundation . . . although . . . an essential component of ethics as an adaptation is that we do believe that ethics does have real foundation (we 'objectify')."[39] In not pretending to draw objective moral norms out of biology, Ruse avoids being swallowed up by any gap between facts and values. Once the strong, naturally selected disposition to believe in some separate domain of values has been described, we've said everything we need to. More ornate explanations fall prey to Ruse's evolutionary version of Occam's Razor. With Ruse's biological explanation in hand, we appreciate that there is no need for any grand metaphysical account of moral objectivity.

Ruse's evolutionary ethic is problematic in two ways. The first question we must ask is whether any power to motivate is retained if Ruse's biological revelations capture all there is to ethics. The second concern is whether he succeeds in quarantining the revisionist implications of evolutionary theory to meta-ethics.

Straight off the bat, it seems that once we realize that ethics really is nothing more than a gene-imposed illusion, we have little reason to act morally. That is, unless we are given substantive moral grounds for taking our genes' interests seriously. Now, there is a sense in which any description of meta-ethical nuts and bolts detracts from the power of normative requirements built out of them. It is hard to match the motivational power of appropriately delivered, but unexplained, ethical commandments. I think that Ruse's evolutionarily driven attack on ethical objectivity has brought him special problems. He tries to make up for any motivational deficit by pointing out that "generally . . . morality is in our own best interests."[40] Is this enough? Take the family of theories that, like Ruse, attempts to get by without any metaphysically awkward ethical objectivity. Social-contract theorists also make much of the idea that overall, morality is in individuals' interests. They have something that Ruse does not, however. A contract, actual or hypothetical, is important because it commits us to morality. Without something to play the role of the contract, what is to stop us from agreeing that morality is generally in our interests but opting out whenever we judge opportune? The hidden stratagems of our genes seem inadequate as commitment devices. Getting away with murder may sometimes be accompanied by psychological trauma in later life, but surely not always.

What impact would this have on the evolutionary environmental ethic? Some will see Wilson's claim as having the reverse consequences to the ones he hopes for. Suppose we reason in the following way. In a pre-enlightened state, our biophilic urges may dispose us to look kindly on arguments for the intrinsic value of nature. These impressions about intrinsic value would be supposed to contain information about some objective feature of the natural world. Science now allows us to see that we are mistaken. Environmental intrinsic value is nothing more than an expedient, gene-imposed illusion. Compare our persistent belief in environmental intrinsic value with phobias. Knowledge that a fear of open spaces is foundationless prods us to overcome it. Why should not the same reasoning hold for the love of nature? Environmentalists will be particularly disturbed that the love of nature exists only because it has contributed to *human* survival and reproduction or, still worse, gene-replication, rather than out of any simple concern for environmental well-being.

So even if the evolutionary ethic explains why our familiar moral concern is patterned as it is, it stands to strip morality of its motivational power.

There is a problem with the general strategy of using evolutionary theory to justify and explain morality as ordinary folk know it. People from different cultures have nearly identical evolutionary histories yet often very different moral codes. Ruse faces up to this cross-cultural variation by pointing to Chomskian grammar. Variation among human languages is compatible with identical underlying structure. Similarly, natural selection might give us the deep structure of ethics. Ruse points out that "particular manifestations of norms may vary according to circumstance, while the underlying structure remains constant."[41]

Recently described biological altruisms make up much of this deep structure of moral theory. Kin-selected altruism describes a disposition to help relatives according to their degree of relatedness. The degree of relatedness will be a measure of genes shared. Reciprocal altruism goes beyond kin. It involves a selected propensity to aid those who are likely to return this help.

Attempts to explain commonsense beliefs about morality raise perennial questions about the relative importance of nature and nurture. Ruse is certainly right that our biology will place limits on ethics. However, it strikes me that these altruisms impose looser constraints on human moralities than those placed on human languages by innate grammar. A very wide range of social arrangements harness our altruistic impulses. It is easy to think of cultures that have made use of our kin-selected urges to convince us that our compatriots, unlike foreigners, are our brothers and sisters. There are very many different forms of social exchange that gain some purchase in our reciprocal altruistic urges. Much of the interesting determining work seems to be done by the local cultural environment. Ruse's evolutionary ethic threatens to collapse into a cultural relativism differing from familiar forms only in its rhetorical focus.

Of course, it may turn out that most cultural graftings onto kin-selected and reciprocal altruistic roots frequently do not increase the fitness of relatives or involve the fair exchange of fitness-boosting efforts. On these occasions, perhaps morality as we know it should give way to a biologically wised-up ethical code. An individual's duties would be determined in relation to his or her potential gene bequests. Substantial measures of kin-selected and reciprocal altruism would be prized because they are part of the optimal gene-bequeathing strategy. Ruse often seems to talk as if he is interested in this type of theory. For example, he takes biological revelations about altruism to undercut "conventional platitudes about obligations to the Third World."[42] Note that this would involve Ruse in doing what he claims he will not do: using evolutionary biology to offer an ethic often significantly different from the widely accepted ethical belief. The gap between facts and values opens anew. It is easy to see how ignoring the needs of the Developing World could be in our interests, but what makes such disregard moral?

Natural Selection of Biophilia or of Bioexploitation?

Whether the facts about selection have, in principle, the right justificatory connection with moral claims is a difficult question. Let's assume that we can find substantive selectional explanations for a wide range of ethical claims and that these explanations offer support to the claims. Even so, I suggest that we are very unlikely to be able to justify any halfway-adequate environmental ethic.

Wilson and Ruse need to map a range of selected urges onto intuitively attractive moral ideas. For example, kin and group selection are invoked to explain our robust belief that our community-mates deserve consideration. This explanation is supposed to provide support for a moral requirement. If no selectionist argument for human tendencies corresponding with seeming norms about nature were available, then our moral instincts in this area would have to have some other origin. Natural selection will not explain them and, therefore, could provide no justification for them.

This spells difficulties for Wilson. There is no match between the pattern of oughts that most would see as appropriate to an environmental ethic and the attitudes to nature that natural selection would encourage. Certain ways of restoring and following our atrophied biophilic urges comport poorly with anyone's conception of a worthwhile environmental ethic.

According to the most widely accepted view, much of human evolution occurred on the African savanna. This makes it the most likely candidate for the Environment of Evolutionary Adaptedness. Wilson notes how this environment has shaped our biophilia.

> Every animal species selects a habitat in which its members gain a favorable mix of security and food. For most of history, human beings lived in tropical and subtropical savanna in East Africa, open country sprinkled with streams and lakes, trees and copses. In similar topography modern peoples choose their residences and design their parks and gardens, if given a free choice. They simulate neither dense jungles, toward which gibbons are drawn, not dry grasslands, preferred by hamadryas baboons. In their gardens they plant trees that resemble the acacias, sterculias, and other native trees of the African savannas.[43]

If the Environment of Evolutionary Adaptedness were composed of African savannas, biophilia will embrace not all of nature but rather a very specific variety. Many humans live in areas whose indigenous ecosystems are very dissimilar to African savannas. Our Pleistocene ancestors would not have enjoyed the rain forest that today's environmentalists enthuse about. Darkness and dense vegetation would have obscured food and concealed predators. It is even possible to imagine appealing to biophilic urges to justify transforming rain forest into pasture land. Pasture land, punctuated by the occasional tree, more closely resembles savanna than does dense forest.

Perhaps Wilson's assumptions about the Environment of Evolutionary Adaptedness are wrong. There is considerable debate as to how long humans and their immediate ancestors have occupied environments outside Africa. Some say one hundred thousand years, others two million years or more. There is further area of uncertainty. How long would humans have to live in an area to acquire psychological adaptations appropriate to its biomes, and what is the pace of evolutionary change in humans?

If humans have lived in environments outside Africa for a relatively long time, and selectionally driven genetic change is quite rapid, then there may be many Environments of Evolutionary Adaptedness. Different biomes will generate their own brands of biophilia.

This outcome would not be of much use to Wilson. It would explain why the preservation of rain forest is a duty of the Yanomamo. But it would be no duty of those whose evolutionary ancestries cannot be traced through rain-forested areas.

Let us put these problems to one side for a moment. There is even some doubt as to whether Wilsonian biophilia could foster respect for environments similar enough to the one in which we evolved. I think that if we ask what type of attitude to nature would have most reliably boosted inclusive fitness in human history, we get a different answer from Wilson. Though we may feel spiritual calm when we are in certain kinds of natural settings, we must ask what the basis of this calm is. The most beloved parts of the Environment of Evolutionary Adaptedness would have been those rich in accessible resources. The biological function of any spiritual claim we feel in natural settings could not have been the preservation of those settings. Rather, its task would more likely have been to draw us toward the resources contained there. If this is so, we should expect that any innate love for the environment would have been that of a greedy consumer rather than of a careful preserver. What might be the significance of the calm that environmentalists enthuse about on confronting wilderness? Those who find a forest clearing first no doubt extract the richest rewards from it. Untouchedness will indicate a probable lack of competition for resources.

Today, human greed for resources has many negative effects on our biological fitness. If environmentalists are right, the destruction of nature and its replacement with a manufactured environment has given us new forms of cancer, crime, and mental illness. However, this type of feedback is a new phenomenon. It is only relatively recently that humans have weighed so heavily on the earth.

We can understand how our selected propensities toward nature might have gotten us into trouble. R. Nesse and G. Williams (1995) discuss a range of diseases that result from a mismatch between the Pleistocene Environment of Evolutionary Adaptedness and modern environments. A predisposition to consume all fat and sugar within reach would have been favored in an environment in which regular supplies of these rich energy sources could not be relied upon. In modern resource-rich environments, however, this predisposition leads to obesity, diabetes, and heart disease. I suspect that if we do have a selected pattern of responses to nature, they will be very much like those we have toward fat and sugar. Attitudes that were well suited to the Environment of Evolutionary Adaptedness are outright pathological today. In this case, we do not end up with bodies ill adapted to survive but with an environment that cannot support us.

During the Pleistocene there were probably not more than one hundred thousand people on the planet. Humans moved about in small groups of perhaps forty to a hundred often closely related individuals. For these few, nature really was an inexhaustible resource. Abuse of the environment would have been unlikely to have been punished as it is today. A filthy campsite could always be abandoned. By the time another band of humans came along, nature would have had time to repair any human-caused damage. If this is the case, efficient bioexploitation, not romantic Wilsonian biophilia, would have been the appropriate trait for nature to hardwire into us. Pleistocene preservationists would have been less efficient at gathering resources than their competitors. This difference in fitness means that there simply could not have been such a thing as a Pleistocene anthropocentric environmental ethic.

Now there are nearly six billion of us, and technology dramatically extends our exploitative power. Nature is unable to bear the burden. Our greedy attitude to nature, like our cravings for sugar and fat, has landed us in a situation in which it could never have placed our ancestors. Any solution to the problem is most likely to require rejection rather than validation of our selected attitudes to nature. We should probably be grateful that centuries of city dwelling have dulled our "biophilic" responses sufficiently to allow us to think in new ways about the environment.

Unfortunately, the arguments discussed in this section have failed to glue moral *oughts* onto environmental *is's*. Achieving this goal will require scrutiny of the concepts that comprise the psychological view about intrinsic value. In the next chapter I begin this task.

CHAPTER 3

⟡

Overlapping Kinds and Value

TWO TYPES OF NATURAL KIND OVERLAP

Folk metaphysics comprises a huge network of largely tacit knowledge about the world. Some parts of this theoretical package guide us in our encounters with other people, other parts in our experiences with nonhuman animals, and still others in our dealings with the inanimate world. For most of human history, the explanatory breadth and predictive power of these folk theories have ensured that they have gone largely unquestioned. The scientific revolution changed all this. Over the past three hundred years science has painted a picture of the world that appears, in places, to differ quite markedly from the ancient folk image. Science's challenge has forced us to question our allegiance to such historically cherished concepts as mind, God, and magic. Only rarely do electron microscopes, gene sequencers, and radio telescopes leave the objects of common sense exactly as they find them. In many cases, the elimination of the folk notion is the only option, while in others, only a radical overhaul offers any hope of vindication.

The aftershocks of science's challenge extend beyond folk metaphysics to morality. Given that commonsense views about how the world ought to be have co-evolved with commonsense ideas about what the world is, it should not be surprising that a significant alteration of the received descriptive views will have implications for moral theory. Successful challenges to folk metaphysics can reshape morality in such a way as to establish the theoretical backdrop for the specific claims about individual and holistic value offered in the remainder of this book.

The following argument takes as its starting point the emphasis on natural kinds in chapter 2's account of moral extensionism. Remember that this cautious revisionist approach does not challenge the deeply entrenched belief that folk psychological notions mark the boundaries of intrinsic value. Rather, ethical extensionists seek to use our commitment to folk psychology as a fulcrum for limited moral revision, urging us to embrace things that science demonstrates to be relevantly similar to the things we are already strongly disposed to value. This strategy has looked to be of only limited use to the environmental ethicist. The focus on psychological kinds promises only to further entrench the contempt for mindless nature. But I will argue for the reverse conclusion. An appropriate consideration of natural kinds linked to value-endowing descriptive terms will expand the category of the intrinsically valuable sufficiently to take in environmental objects.

I begin with a problem for natural-kind-assisted ethical extensionism. In the last chapter I assumed that the kinds associated with sentience or rationality would be easily recognizable and separable from other kinds, and hence that it would not be difficult to see where to expand and where to contract the category of intrinsically valuable things. Is this assumption correct? Though we may have all the relevant information about natural kinds and about a given value-endowing psychological notion, we will find the choice of kind to guide our revisionist efforts underdetermined.

To find principles that could guide us in scientifically tidying up moral categories, we need to look into the relationship between ordinary language—descriptive notions and natural kinds. This will require a detour though semantic theory.

Semantic theorists have long pondered the relationship between ordinary-language names and natural kinds, seeking to find out what might bind a given name to a specific kind. Now, we have already found grounds independent of any theory of meaning for thinking that certain natural-kind boundaries are morally interesting. Even if semantic experts tell us that no commonsense psychological name means or refers to any natural kind, we still have reason to adjust the concern associated with the commonsense psychological name so that it tracks the relevant psychological natural kind.

Nevertheless, a discussion of semantic issues can be useful. Semantic theorists are well placed to point us toward the resources associated with folk psychological names that might guide us in our moral revising, and my demonstration of a morally troubling kind-indeterminacy will proceed by way of the semantic theorist's problem of kind-selection.

Kind Indeterminacy and the Classical View of Concepts

There are two widely discussed theories that promise to explain how to map a term onto a particular kind. These are the *description theory of reference* and the *causal theory of reference*. According to the first, descriptions associated with a term account for the meaning of that term. When its meaning "contains" what is deemed to be essential to a natural kind, it refers to that kind. For example, if our analysis of the meaning of *gold* uncovers "element with atomic number 79," then we can straightforwardly trace the term to the kind.

Problems arise when we ask what does the job of associating descriptions with terms. On most accounts, everyday semantic intuitions, rather than the views of experts, are the associative glue. This leads to difficulties. Very rarely will descriptions commonsensically tied to a term lead unambiguously to a single natural kind.

To better appreciate the implications, I turn to an area of discourse filled with terms likely to pose problems similar to those posed by value-endowing folk psychological notions. There has been some debate over whether psychological illnesses fall into natural kinds.[1] As with other psychological kinds, we will almost certainly be making a mistake if we say that the brains of all manic depressives share some microstructural property accounting for their symptoms. However, it does seem possible that the diagnoses and predictions of a perhaps future psychiatry will be grounded in some pattern of objective samenesses. In chapter 2 I outlined a liberal account of natural kinds that would cover such cases. According to such an account, objective similarities fit to support explanation and prediction may be found at higher rather than microstructural levels in nature.

If psychiatric kinds exist, many popular beliefs will be poor guides to them. Take the beliefs guiding the current usage of terms like *schizophrenia* and *depression*. It seems likely that these associated descriptions—"acts unhappy," "sometimes makes socially inappropriate comments," "takes him- or herself to have special powers"—will not allow us to pick out any one psychiatric kind. Matters become more fraught once we add mistaken beliefs to the descriptive mix. It is likely that some descriptions widely allied to terms of psychiatric illness, such as "always dangerous to children," will be true of no kind.

What goes for concepts of psychiatric illness stands also for value-endowing psychological notions. Predictive and explanatory successes may inspire confidence that there is some pattern of objective samenesses in the vicinity of folk psychological concepts. However, we have little reason to think that the sets of beliefs surrounding folk psychological concepts will relate to kinds in a one-to-one fashion. Further, it is surely the case that some widely accepted beliefs about psychological categories will end up misdescribing kinds.

Before saying more about how well the description theory fills our moral theoretic role, I want to describe and examine another semantic theory. According to the causal theory of reference, first defended by Saul Kripke (1980) and Hilary Putnam (1979), the meaning of a term is determined by a causal chain linking it to a particular individual or property. This causal chain comprises a reference-fixing component and a transmission-of-reference component. The reference-fixing story tells us how the causal chain attaches to the world. According to the *Oxford English Dictionary*, the terms *sentience* and *sentient* originated in the 1600s. We can imagine some seventeenth-century figure introducing the term after wondering what sets humans apart from chairs and tables. The dubbing situation causally connects a sentience kind or category with *sentience*, and Singer's and our usage lies at the far end of one of the many causal tentacles emanating from this original fixation of reference. For the causal theorist, it will not matter that these early referrers had mistaken views about sentience, perhaps supposing it to be a nonphysical state possessed exclusively by humans, so long as the dubbing ceremony causally binds it to some pattern of objective samenesses.

The Problems of Kind Overlap for Both Causal and Description Theories

Of the two accounts, the causal theory seems more suited to the needs of ethical extensionists. While everyday semantic intuition may equivocate, causal paths seem to emanate straightforwardly from natural kinds. However, the causal theory proves something of a disappointment to the ethical extensionist.

In their discussion of the causal theory, Michael Devitt and Kim Sterelny (1987) describe the *qua* problem, one manifestation of which concerns kind overlap. Their example is the attempted reference to a particular native Australian animal: "Any sample of a natural kind is likely to be a sample of many natural kinds; for example, the sample is not only an echidna, but also a monotreme, a mammal, a vertebrate, and so on. In virtue of what is the grounding of it in *qua* member of one natural kind and not another?"[2]

Information about causal chains fails to single out one of these kinds as the appropriate referent for a term. Given that *echidna* attaches causally to each of the kinds listed by Devitt and Sterelny, what grounds do we have to privilege any of them?

The moral theorist stands to be as perplexed about kind overlaps as the semanticist. Many natural kinds, each with very different boundaries, run through human heads. Moving down from the higher to the lower levels, we have the kinds of psychology, neuroscience, chemistry, and physics. Without a solution to the problem of kind-overlap, we will not be able to decide which kind boundaries are to be morally significant.

Now, more restrictive views of natural kinds would certainly mollify matters. Were we using microstructure as the marker of kindhood, we could immediately dismiss from consideration many candidates for reference in the Devitt and Sterelny story, certainly all the candidates they mention, leaving only the physico-chemical kinds. This option is not open to us. To allow for the psychological kinds necessary for moral revision, we have abandoned sameness of microstructure as the unique criterion of kind-hood. It is precisely this abandonment that makes problems of kind-overlap ubiquitous.

Devitt and Sterelny take the *qua* problem to derail pure causal theories of reference and go on to urge a compromise between causal and description theories. Such a compromise would give access to information about the descriptions that the referrer associates with the term. Though causal theorists often point out that meaning intuitions have no special authority in sketching the boundaries of natural kinds, intuition may be all that we can appeal to in situations in which there are more than one candidate kind. It is easy to anticipate how this might work. Though their boundaries are different, the kind made up of organic compounds and the kind comprising hearts have members in common. Commonsense views about the everyday concept *heart*—that it is a blood-pumping muscle, can be damaged by cholesterol buildup, and so on—help show us which of these two kinds will be uncovered by the theoretical deepening of the concept. If we accept the psychological view as our moral starting point, we will use the intuitive theory surrounding folk psychological notions to guide us toward the kinds that make two otherwise dissimilar objects morally similar.

Matters are not quite so simple as this. Both the description theory and the causal-descriptive hybrid face problems of kind overlap. Consider an example from biology. The upshot of twenty years of debate is that we have at least two options when deciding how to naturalize the notion of biological function, or biofunction. The etiological theory defines function in terms of selection history. Here is Karen Neander's statement of the theory: "It is a/the proper function of an item (X) of an organism (O) to do that which items of X's type did to contribute to the inclusive fitness of O's ancestors, and which caused the genotype, of which X is the phenotypic expression, to be selected by natural selection."[3] So on this theory, my kidney has the proper function of purifying blood because things of the kidney's type contributed to the inclusive fitness of my ancestors, thereby causing the genes underlying the kidney to get selected, by purifying blood.

Opposed to the selectional etiological theories are the goal theories of Christopher Boorse (1984) and John Bigelow and Robert Pargetter (1987, 1990), among others. These theories assign functions to organs based on their propensity to contribute to some future goal, or goals, of its possessor. The most recent version of the goal theory is the propensity account argued for by Bigelow and Pargetter. According to them, "something has a (biological) function just when it confers a survival-enhancing propensity on a creature that possesses it."[4]

This survival-enhancing propensity is to be measured in the creature's natural habitat. The propensity theorist says that a heart possesses the function of pumping blood because it confers a survival-enhancing propensity on its possessor by pumping blood.

Both etiological and goal theories are constructed out of terms eminently suitable to describe natural kinds. "Natural selection" and "survival enhancement" figure in our most up-to-date biological theories. So which kind is the naturalization of the ordinary-language notion? Etiological and goal accounts have their own particular intuitive weaknesses. Most strikingly, the etiological account faces the objection that it seems at least possible that there could be functional yet entirely ahistorical entities. Take Donald Davidson's story about Swampman:

> Suppose lightning strikes a dead tree in a swamp; I am standing nearby. My body is reduced to its elements, while entirely by coincidence (and out of different molecules) the tree is turned to my physical replica. My replica, The Swampman, moves exactly as I did; according to its nature it departs the swamp, encounters and seems to recognize my friends, and appears to return their greetings in English. . . . No one can tell the difference.[5]

What should the etiological theorist say about Swampman? As it is a completely ahistorical entity, there seems no alternative but to deny it any biofunctioning parts at all. Somewhat surprisingly, Swampman will have no heart that functions to pump blood; in fact, as *blood* is a biofunctional concept, it will possess no blood. While I am an incredibly complex collection of millions of functioning parts, Swampman comes as close to possessing functioning parts as a rock at the bottom of the ocean.[6]

Goal theories have their own faults. Millikan and Neander both emphasize the benefits of historically determined norms. Alternatives to the etiological theory account for functions by choosing from the class of a biological object's *actual* dispositions; these dispositions might be those that are most statistically frequent in some accounts, or those that contribute to some given goal in others. However, it is part of common knowledge about function that biological objects are typed on the basis of what they are *supposed* to do, rather than what they actually do. Millikan's striking example of the spermatozoon illustrates very well the superiority of historical notions in accounting for these norms. The overwhelming majority of spermatozoa have no disposition to fertilize any ova unless dispositions are calculated against extremely biologically improbable backgrounds. Yet we still say that the function of sperm is egg fertilization. The etiological theorist recognizes that the sperm has this function because egg-fertilization is what it is selected for.

I will say no more about functions. Those who doubt the difficulty of the problem will be able to trace the debate through scores of journal articles.[7]

Descriptive and Metaphysical Kind Overlaps

We can now identify two types of kind overlap. Each of these is associated with one of the earlier discussed ways of binding terms to natural kinds. *Metaphysical kind overlaps* are a concern for the causal theorist. They occur when different natural kinds cohabit in a given object or set of objects, such as in Devitt and Sterelny's echidna. A large number of metaphysical overlaps should not surprise us. As we have noted, they result from the abandonment of microstructure as the unique hallmark of kindhood. The stuff that makes up a human brain is the proper object of study for psychologists, neuroscientists, evolutionary theorists, geneticists, and chemists. Workers in each of these areas examine the same object, but their investigations will track different kinds.

Descriptive kind overlaps plague description theorists in their efforts to single out kinds. Here more than one kind is a candidate for a given commonsense term even after we have appealed to semantic intuitions. Associated descriptions fail to lead unambiguously to a kind.

The notions of metaphysical overlap and descriptive overlap themselves overlap. Both backward- and forward-looking function-kinds are to be found in the same bird's wing. The wing has been selected for certain effects but also has a propensity to be selected for certain effects. Descriptively overlapping kinds need not always metaphysically overlap, however. We've seen two biological naturalizations of the commonsense notion of function. Engineers and architects use the term *function* to describe a class of objects completely distinct from those that interest biologists. When setting out to understand the functions of bridge or house parts, we look to designers' intentions. Thus, in competing for the same commonsense descriptive term, intended-functions and both types of biofunction descriptively overlap. But they will not metaphysically overlap. Things that get their functions by virtue of a certain design history will not also get their functions by virtue of selection history. In transforming a biological object into food or house parts, we normally consider the design-functions to supplant the natural ones, rather than to coexist with them.

WHAT TO DO ABOUT KIND OVERLAPS

Now that I've described metaphysical- and descriptive-kind overlaps, I want to say how those engaged in both descriptive and prescriptive endeavors should respond to them.

Neither type of natural-kind overlap presents any lasting difficulty to those seeking to describe the world, so long as explanatory goals are kept in mind. To return to our earlier example, the semantic confusion about biofunction is no obstacle to science. The modern biologist faces requirements that could not possibly have been foreseen when the forces that shaped commonsense intuition about function were in operation. Freed from the obligation to find the unique kind meant by ordinary folk when they use the word *function*, our emphasis in naturalizing can be different. Rather than seeking to explain a familiar notion in naturalistic terms, we set out to see how the notion together with its accompanying theory can assist a given naturalistic enterprise. Though the theory accompanying the commonsense notion of function should be useful, resulting, as it does, from millennia of folk experimentation with functions, it is not immune from error. We take the term, retain some of its associated folk theory, and supplement it so as to satisfy certain domain-specific theoretical requirements.

When a paleontologist asks what the function of the hard crest on the heads of some duckbill dinosaurs was, this question is best treated etiologically. She or he is asking what properties of the crests have accounted for their existence. If it turns out that they were merely side effects of other features and were not directly explainable by reference to natural selection, the paleontologist no doubt would say that they were functionless, regardless of how happy they may have, on occasion, made individual duckbills. This all makes sense when we think about what paleontologists do. When they talk about functions, their concern is to discover which facts, if any, about a certain piece of morphology, such as the duckbill crest, explain why it continued to appear in the fossil record over millions of years. The "organs" of a fossil Swampduckbill would for a paleontologist count as functionless because these features got there by chance, rather than through the action of selective forces. Its "crest" would belong to an entirely different category from the crests of normal duckbills. So the usage of paleontologists requires a historical notion.

For an illuminating contrast, take the case of a medical researcher. Some biologists believe a few old-age ailments in humans are the result of a kind of built-in obsolescence.[8] They hypothesize the existence of biological mechanisms selected for their propensity to cull out members of the population who are past reproductive age and are no longer capable of effectively contributing to their relatives or communities. Even if this does not occur, and these mechanisms do not exist, they are certainly within the realm of biological possibility. Imagine that this selected-for phenomenon manifests itself in accelerated tissue breakdown after a certain age. In reversing this breakdown, a medical researcher seeks to interfere with selected function. Nevertheless, he or she might likely see the reversal as *restoring* rather than *impeding* function. If so, the notion to which the medical researcher will want to appeal will be some kind of goal account, rather than the etiological theory. It follows that here function will depend on a disposition to contribute to individual fitness or well-being.

We need some new terminology to describe these descriptively overlapping kinds. Where we can map an ordinary-language notion clearly and exactly onto one kind, we talk of that kind as a reductive naturalization of the term. Where there is no unique kind, but instead many kinds, I will use the label *plausible naturalization.* If a folk notion appropriately pruned back then augmented can often prove useful in different theoretical contexts, then we will have many different plausible naturalizations.

Being a plausible naturalization is most often a matter of degree. We can place natural kinds on a continuum that measures how much of the folk theory attaching to a commonsense notion is profitably used to describe them. Plausible naturalizations fruitfully appeal to a substantial chunk of theory in delimiting kinds. At the other end of the spectrum, some conceptual descendants may be sufficiently far from the ordinary-language notion that the kinds picked out will seem very counterintuitive referents for the ancestor notion. A kind can be a perfectly respectable plausible naturalization of a commonsense concept while being just hopeless as a naturalistic reduction of that same concept. While in these cases we should expect that intuitive theory offers fewer clues to kind boundaries, this greater distance need not gainsay the role the ancestor notion has played in the formation of the new concept, nor the theoretical utility of this new concept.

Biologists often use apparently mind-invoking language, such as talking about altruism and selfishness to explain evolutionary theory. Richard Dawkins (1989, 1990) uses the term *selfish* to name a propensity of genes, under a natural selective regime, to produce phenotypic characters that boost their own survival chances. Despite the concern of some critics, there seems good reason to think that this property of genes might be a plausible naturalization of "selfishness." Though very far from the ordinary-language notion, talk of selfishness seems relevant in this evolutionary biological domain. This relevance is premised on the isolation of some significant biological regularity.

Investigators whose primary interest is meaning will be wary of sticking with commonsense terms for sufficiently distant plausible naturalizations. Remember, however, that meanings are interesting to us insofar as they provide the raw materials for accounts of natural kinds. The boundaries of certain natural kinds show how moral concern ought to be distributed regardless of what semantic experts tell us about *consciousness, belief,* or *selfishness*. This change in focus enables us to put to one side much of the philosophical baggage traditionally associated with semantics. Take the long-maintained distinction between essential and accidental properties. The traditional task of a definition is to isolate essential properties, those that a given object or type of object could not lose and still be the same thing, or type of thing. Accidental properties, on the other hand, can be lost without compromising identity. Gold is both yellow and the element with atomic number 79. Yellowness could be lost by a golden thing without stopping it from being gold. It is, therefore, an accidental property of gold. In contrast, to lose the property of being the element with atomic number 79 is to stop being gold.

It is certainly the case that a plausible naturalization of selfishness that does the work Dawkins requires of it will be far from the commonsense concept. Hostility among philosophers more or less wedded to the program of ordinary-language philosophy is evidence for this difference. Take the criticisms of Mary Midgely. She begins an attack on Dawkins with the claim that "genes cannot be selfish or unselfish, any more than atoms can be jealous, elephants abstract or biscuits teleological."[9] Midgely observes that selfishness can be applied only to a being capable of thought and conscious motives. Such properties are essential to the notion: "One cannot speak even of 'unthinking selfishness' in beings incapable of the thought in question. Most selective competition does not require competitive motives, nor any sort of motive involving calculation of consequences. . . . Nobody attributes selfish planning to a paramecium. What, then, can Dawkins mean by attributing it to a gene?"[10] Midgely is far from showing, as she hopes, that Dawkins's use of the word *selfish* is not perfectly appropriate within the evolutionary biological domain. What she does succeed in demonstrating is that Dawkins's notion is rather distant from the commonsense notion. To pick out the relevant natural kind, we need to strip away a considerable amount of the ordinary-language notion's accompanying theory.

The first thing to note is that the intuitions that allow semantic theorists to divide properties into the important essential and less-important accidental are vestiges of yesteryear's descriptive theories. For the natural kinds–focused account that I favor, specialists in a given domain, equipped with an up-to-date understanding of relevant objects, determine which theory to retain and which to dump. As science uncovers unexpected natural similarities and differences among objects apparently familiar to folk categorizers, established boundaries between the essential and the accidental become warped.

We can replace the essential/accidental distinction with another one. An interest in explanation will allow us to divide generalizations surrounding a term into *core* and *peripheral*. This distinction is made in terms of the range of kinds for which a given generalization is true. Peripheral generalizations will tend to be true for a large number of kinds. Among them will be those that encode information about the laws of physics governing every material object. Core generalizations will begin to carry information about a smaller range of objects. They help us see what marks this kind off from others and should, therefore, be given more weight when deciding how plausible a given kind is.

Interest in core and peripheral generalizations allows us to distinguish scopes of plausibility. Some naturalizations are more *globally* plausible, or intuitive, than others. Global plausibility is measured against *all* the kinds that we could choose to describe by using a particular term. The kind that comes closest of all to meeting the intuitive requirements will be the most globally plausible. Contrast this with *local* plausibility, assessed within a given explanatory domain. To make a judgment about local plausibility, we will line up all of the kinds relevant to a given area, say biology or a subdiscipline of biology. Global plausibility tends not to be a big issue when we assess the use of terms within specified theoretical domains. To perform a service within a given domain, a commonsense term need only be matched with a kind that has the greatest local plausibility. A globally rather implausible kind might be the most plausible within its domain. Genic selfishness may be the most locally plausible natural kind. However, it is globally implausible.

DESCRIPTIVE OVERLAPS AND MORALITY

How should moral theorists respond to each type of natural-kind overlap? I start with the less problematic metaphysical overlaps. Some of the kinds found in us are also to be found in a dead human body; others will be shared with an intelligent extraterrestrial. The psychological view together with semantic intuitions will guide us toward kinds shared with the extraterrestrial, not with the dead human body. It is psychologicalness, not carbon-ness, that makes something worthy of the same manner of moral consideration that we accord ourselves.

Cases of descriptive-kind overlap are potentially more troubling for moral theory. Again, there is likely to be an array of more or less plausible naturalizations for most value-endowing psychological notions, and when we have multiple plausible naturalizations, we need to decide which kind should act as the platform for extending moral value. Take the concept of intelligence. There are many different answers to the question "What is it for something to be intelligent?" We have the famous behaviorist notion due to Alan Turing, which makes central the production of sophisticated verbal behavior over a sufficiently long period.[11] A battery of criticisms demonstrates the gap between this and our familiar concept of intelligence. Objectors urge that a device can match perceptual input with apparently appropriate behavior in patently unintelligent ways.[12] The commonsense concept requires sophisticated internal processing, and here is the prime intuitive failing of the behaviorist accounts of intelligence. We must remember that regardless of how many folk requirements it meets, this notion has played an important role as a research target in artificial intelligence.

We can meet more intuitive requirements by providing more detail about internal processing. A commonsense functionalist account specifies internal states arranged as prescribed by the host of folk psychological platitudes.[13] Extremely powerful input/output devices fail to meet this requirement.

Beyond this, we can identify a broader, ethologically helpful use of the label *intelligent* for systems that form goals and act in such a way that satisfies these goals while still allowing for perturbations in the environment. Some predators are capable of responding to only a very narrow range of prey stimuli. Others can identify their quarry against an array of backgrounds and respond appropriately to a number of evasive strategies.[14] These predators do not talk, and it is unlikely that they qualify for the full folk psychological notion of intelligence.[15] Neither of the first two definitions, then, will count animals as intelligent. However, this third account includes nonhumans capable of sufficiently diverse and sophisticated behavior.

Intuitive deficiencies do not prevent these notions from being theoretically useful plausible naturalizations of *intelligence*. We've seen that some conceptual distance need not be an obstacle to theoretical utility. Ideally, each concept will borrow from its commonsense ancestor so as to better describe appropriate kinds.

Similar conclusions may be drawn about other psychological notions. A great deal of effort has been invested in working out in naturalistic terms what it might be for a mental state or representation to have a given meaning. Stephen Stich points to the "dizzying range" of attempted naturalizations of "representation": "There are causal co-variational theories, teleological theories, functional role theories, and theories inspired by the causal theory of reference. There are single factor theories, multiple factor theories, narrow theories, wide theories, and a profusion of variations on all of these themes."[16] I cannot go into detail about the naturalistic approach to meaning here. However, there is an approach consistent with what I have already said about "selfishness" and "intelligence." The plausible naturalizer is not so concerned with closeness or distance from the commonsense concept of representation. Rather, the requirement will be for the proponent of each candidate account to show how the naturalization can satisfy a particular theoretical need. "Content" is clearly folk theoretic in origin. However, different types of content might best serve connectionists, computationalists, ethologists, and so on.[17]

Significantly for us, *intelligence* and *representation* are morally loaded terms. According to the traditional view, an object can be eligible for intrinsic value by virtue of representing its environment, or being intelligent. So we need to ask what approach the moral philosopher should take to plausible naturalizations of folk psychological notions.

If we are interested only in describing the natural world, the relationship of terms to our traditional notions may ultimately matter little. Traditional notions and associated theory serve as a pool for the construction of new terms. However, once we have the term, facts about its conceptual ancestry do not bear on how well it helps the relevant specialist describe the natural world. Realists about the objects of scientific inquiry will be interested in how much hitherto anomalous world-structure the new term helps illuminate, and facts about the history of usage of the folk notion will eventually drop out of the picture.

Things are more complicated for the ethicist. According to most accounts, the moral enterprise is community oriented in a way that science is not. Moral philosophers tend not to see themselves as describing some realm of objective moral reality complete with its own microstructure and lawlike generalizations. Moral rightness is often tied to community views, be they actual or ideal. Communities are guided by commonsense notions commonsensically interpreted, not by a scientist's terms of art. This means that commonsense views about the properties picked out by the ordinary-language notions matter for the moral philosopher in a way that they do not for the scientist.

Assigning Value to Plausible Naturalizations

Two approaches to the plurality of plausible naturalizations corresponding to a morally loaded commonsense notion suggest themselves.

The most obvious approach would be to allocate all moral considerability carried by the old term to what we adjudge is the *most* plausible naturalization. Though it is unlikely that there will be an exact fit, we can use conceptual analysis to gauge distance from the ancestral notion. The fewer counterintuitive consequences that arise from a candidate definition, the more globally plausible it is.

When our aim is to describe the natural world, local rather than global plausibility tends to matter more. Our capacity to apportion theoretical interests among domains means that we need not hunt out a single notion. Global plausibility becomes more important for moral theory, as here we do not have science's division of kinds by theoretical domain. Objects best described by specialists in different disciplines will often need to be morally weighed against one another. Such a need arises when we must decide how many objects falling under one kind-term we need to sacrifice to conserve a certain number of objects falling under another.

This might give us reason to think moral theorists should be interested only in the naturalization occupying the global maximum. In my view, except in the odd cases in which a folk notion together with theory exactly picks out a kind, it would be a mistake to assign all the ancestral moral load to this one notion. We have seen that naturalizations, whether they are the most plausible or not, carry over chunks of the descriptive theory normally accompanying the ordinary-language ancestor. The utility of this theory in the new context does not demand uniqueness. It may be that much folk theory can be profitably used in one domain, but this does not preclude less of that same theory from being profitably used elsewhere.

I think that we should treat the moral generalizations adhering to the original term in a similar fashion. Again, there is no requirement for uniqueness. The fact that there is a most globally plausible naturalization does not rule out transference to less plausible naturalizations. Value-endowing descriptions may help describe a kind regardless of how well they apply to other kinds.

There is another important question. What is transferred from the ancestor notion to a plausible naturalization? Here there is a significant disanalogy with science. With descriptive theory, we can often be quite precise about which parts of theory apply to a plausible naturalization. The physical makeup of the kind means that a given chunk of descriptive theory is true of the kind while another chunk is false. Properties commonsensically associated with a notion can come to be neatly explained by reference to specific attributes of the natural kind. Why is steel a shiny white color? Because its surface properties reflect all wavelengths of light equally well. Why does steel conduct electricity? Because it has very mobile electrons. We cannot make similar claims when it comes to normative theory. Science may one day tell us about the many natural kinds underlying rationality or consciousness. However, these scientific accounts are highly unlikely to break down into component parts in any way that could be mapped onto our preferred decomposition of the moral theories surrounding rationality or consciousness. The relationship between the moral story and a given scientific story will be whole-whole rather than part-part.

What is transferred, I claim, are not specific normative generalizations, but instead degrees of value carried by the ancestor notion.

This leads to the right way for moral theory to treat multiple naturalizations. I propose that the moral significance that traditional morality attaches to the commonsense notions be distributed among descendent plausible naturalizations. The closer the notion picking out a plausible naturalization is to the ordinary-language concept, the greater the degree of moral weight that will be inherited from the ancestor notion. Thought experiments whose goal is to measure conceptual distance should allow us to distribute the moral value attaching to the scientifically dubious but morally clear-cut ancestor notion among its natural-kind-respecting descendants.

An example will show how a strategy that embodies this principle differs from the traditional approach. There has been considerable debate over whether a sophisticated computer might count as intelligent in itself, rather than as just an extension of a designer's intelligence. The traditional approach would be to await the resolution of this debate before the moral status of the future computer could be decided. If it turns out that no set of internal states could ever make computers genuine referents for the term *intelligent,* then none of the moral value that beings are supposed to have by virtue of being intelligent could ever attach to them.

The strategy I urge dodges these semantic strictures. Though a computer may never possess states that help it satisfy the rich folk conception of intelligence, it can, perhaps by passing the Turing test, possess states that count as a plausible naturalization of intelligence. Putative intuitive counterexamples should then probe conceptions of intelligence that would count the computer in. Responses to thought experiments will help us discover what discount rate to apply to the value carried by the familiar notion.

I have argued that normative generalizations carry over together with descriptive generalizations to plausible naturalizations of commonsense concepts such as *intelligence.* The objector will claim that plausibly naturalized intelligence turns out to have very little to do with real intelligence. Objects that are a sufficient distance from the commonsense notion to miss out on what semantic experts deem the "essential properties" of intelligence will be nothing more than "as if" intelligent, or intelligent by analogy. In these cases, the language of analogical extension would seem far more appropriate than any talk of naturalization. Unlike with "the real McCoy" properties, normative or descriptive generalizations may not carry over. As cockroaches will be only "as if" intelligent, they cannot inherit any value attaching to the familiar notion.

Now, a point similar to the above can be made within the approach that I am advocating. Failure to possess what common sense deems to be essential properties may be an indicator that many of the core generalizations are not carried over. However, it can certainly be the case that significant descriptive theory carries over to objects that semantics experts judge to be missing an essential property. Imagine that after centuries of thought experiments we decide that the possession of folk psychological states is a nonnegotiable component of the ordinary-language concept of intelligence, and that languageless nonhumans do not possess folk psychological states. Regardless of this important conceptual advance, generalizations close to the semantic core of the concept may still be true of nonhumans. Take the claim that intelligent things often respond to their environments in ways that satisfy needs or goals. Even when we have decided that language is essential to intelligence, we will still consider this an important component of theory about intelligence. Non–folk psychological cockroaches do not act merely *as if* they are sensitive to ecologically salient properties, they genuinely are.

My suggestion is that when sufficiently many descriptive generalizations apply to an object, associated normative generalizations begin also to apply. Lack of an "essential" property is not an absolute barrier to value. Of course, properties that we adjudge essential may tend to be central rather than peripheral. Kinds lacking these properties will tend to be rather implausible and, therefore, will inherit rather little value.

A note of caution. Success requires that we make a good job of the conceptual stage of our enterprise. We must decide exactly how normativity is distributed among a given set of descriptive notions. Some notions may bring with them a certain type of consideration due to a close association with other descriptive notions, and it may be that we can find naturalizations of the first notion in a given system without finding naturalizations of the latter notions. In this case, the value that attaches to the first notion by virtue of connection to the second ought not to be transferred. An example to be discussed later plays intuitions about the moral weight of intentional and phenomenological states against one another. As most intelligent things we meet can also feel pain, we are accustomed to treating them accordingly. However, we may judge that a system is a plausible naturalization of *intelligence* without finding in it naturalizations of the relevant phenomenal states. The right reaction to this discovery will be to appropriately discount any specifically intelligence-related value without carrying over the respect associated with the capacity to feel.

COMBINING DESCRIPTIVE AND METAPHYSICAL KIND OVERLAPS TO UNEARTH ENVIRONMENTAL VALUE

I have suggested how the moral theorist should approach kind overlaps of both the metaphysical and descriptive varieties. In the case of metaphysical overlaps, we must extend moral consideration to objects sharing the appropriate kinds. I have also urged that we approach descriptive-kind overlaps by deciding the degree of plausibility of a given kind and assigning value accordingly. Now how can the environmental ethicist take advantage of both metaphysical- and descriptive-kind overlaps? The kinds that ecologists describe will metaphysically overlap with morally important kinds. Why do I say that the kinds metaphysically overlapping with ecological kinds are morally important? Because they are plausible naturalizations of value-endowing folk-psychological notions.

So we start with a scientific account of a certain set of environmental objects. We then scour this account for clues to the existence of kinds that descriptively overlap with folk psychological kinds. In succeeding, we show these environmental objects to be intrinsically valuable.

Earlier in this chapter I described Dawkins's discovery of a plausible naturalization of selfishness in genes. His vindication of this psychological description should permit us to use our principle of intuitive closeness to assign moral weight to genes.

There are a number of reasons for thinking that any moral considerability assigned to genes will be very small indeed. First, it is doubtful that selfishness, in isolation, is one of the ordinary psychological notions to which much positive moral significance is attached. There is also the question of intuitive distance from the commonsense notion. Remember that this determines the proportion of the original moral load that is inherited. We saw that genic selfishness is rather distant from commonsense selfishness. Though occupying a local plausibility maximum—it is the appropriate kind within evolutionary biology—genic selfishness is globally implausible. We should conclude that Dawkins can use the word "selfish" to describe genes without conferring any significant value on them whatsoever.

Here is another candidate environmental-psychological kind. We have seen that there might be more than one plausible naturalization of intelligence. Jonathan Schull (1990) outlines what might be another. He argues that species are intelligent. Note that Schull is interested not in species at given times, but species lineages extending over time. He says,

> . . . plant and animal species process information via multiple nested levels of variation and selection in a manner that is surprisingly similar to what must go on in intelligent animals. As biological entities, and as processors of information, plant and animal species are no less complicated than, say, monkeys. Their adaptive achievements (the brilliant design and exquisite production of biological organisms) are no less impressive, and certainly rival those of the animal and electronic systems to which the term "intelligence" is routinely . . . applied today.[18]

The gene pool of a species-lineage is informationally sensitive across time to its environment. Information allows the lineage to adapt to this environment. A range of possible solutions to problems—such as predation or competition from another species for a resource—is generated. Selection rewards success as some strategies out-reproduce others, resulting in a greater adaptive fit. Schull places this account alongside the structurally similar story about individual organisms' intelligence. Here we have individuals drawing on a store of information to form hypotheses about their environments. These hypotheses are then tested, and those that are inadequate discarded; those that lead to success are reinforced. Both processes tend to lead to "behavior" appropriate to a given environment.

Let's assume that the commonsense notion of intelligence excludes species. They are also dubious referents for the other notions outlined earlier in this chapter. Little of folk psychology can be true of species, and they do not talk. Nevertheless, we may well have a kind that certain specialists find expedient to label intelligent. If this is the case, then we may have yet another plausible naturalization of that notion. Sufficiently savvy intelligence-talk will allow scientists to generate hypotheses that help explain species' role in the natural world.

To establish whether any moral considerations are involved, we must treat intuitive objections to this account in the above-described fashion. Moral value attaches most straightforwardly to the commonsense notion. We probe conceptual descendants of the original notion with intuitive counterexamples.

We must exercise caution in deciding how value is to be transferred to this particular descendant. Some value-bearing notions allied to intelligence may carry over with a similar rate of discount to intelligence. Others may be less smoothly transferable. Destruction of a subspecies of the black stilt may cause pain to individuals, but the idea of any species-pain seems a real affront to common sense. Of course, it is not out of the question that a plausible naturalization of pain might be found in species. Pain is likely to be at least partially functionally definable. This functional characterization might help us understand what goes on in species when they are threatened with extinction, for example. We must then decide whether this functional characterization captures the associated quale and what proportion of the moral value linked to pain attaches most directly to these qualitative aspects.[19]

Once we have vindicated the ascription of the term *intelligence*, the way is open to attempt to locate in species the plausible naturalizations of a range of intentional notions. An intelligent species may have beliefs about its environment and may be capable of forming desires to survive.

Despite all these potential areas of investigation, I do not think that Schull's argument points us toward the best path to species value. In showing that the term *species intelligence* isolates a natural kind, we also demonstrate that the kind in question inherits some of the value associated with our commonsense notion. Sadly, species intelligence turns out to be rather implausible. This implausibility will arise in the case of any holistic naturalization of properties normally realized by individuals. One way of expressing this worry points to a supposed conceptual link between intelligence and the possession of qualitative states:

> The primal intuition on which our notion of intelligence draws is subjective (and is hence wedded, for better or for worse, to the mind body problem). . . . Is a nation or corporation or a species the kind of thing that can, like the compact symbiotic population of living cells we are, have a toothache? If not, is it really the kind of thing that can be intelligent?[20]

If, as the editorial commentary on Schull's paper suggests, our commonsense notion of intelligence is intimately linked to the possession of qualitative states, then no qualia, no intelligence. Well-known objections to functionalist accounts of mind have centered on spatially distributed entities. Ned Block (1991) has suggested that the interactions of the billion inhabitants of China might instantiate the preferred functionalist theory of mind. Despite this, he claims that it is very unlikely that such a thing could have qualia.

So, spatial scatteredness causes problems of plausibility. There is also the matter of temporal distribution. Species respond to challenges posed by their environments intergenerationally. Now this does not always take large amounts of time. For example, the replication rate of HIV is usually sufficient to generate billions of new HIV particles in a single day. However, in many cases the intelligent behavior of species will be for us almost imperceptibly slow, best tracked in the fossil record.

Both spatial and temporal distribution make for thought experiments that demonstrate a divide between commonsense intelligence and Schullian intelligence.In the next chapters I will examine a more serious candidate for a kind that descriptively overlaps with value-laden folk psychological notions and metaphysically overlaps with things in the environment. To conclude this chapter I will summarize my approach to the revision of the psychological view about intrinsic value.

The scientific investigation of folk psychological notions stands to have one of three outcomes. We might join Patricia Churchland and Paul Churchland in deciding that folk psychology comes nowhere near isolating any kind or regularity in nature. Or if we find a close match between the folk-theoretic package and a single kind or set of natural regularities, we can look upon folk psychological states as straightforwardly naturalistically vindicated. There is a third, more likely, alternative. We might decide that there is more than one kind that we can illuminate by using the folk package as a foundation. Folk psychological theory will not guide us with respect to a unique kind or regularity. Instead, different chunks of it, appropriately augmented, will inform us about the boundaries of a range of kinds.

The three possible fates for value-anchoring notions have differing implications for ethics. If folk psychological states are to be eliminated, we would then need to undertake the tough task of finding a new boundary between the morally considerable and the morally inconsiderable. If we can find unique perfect deservers of folk psychological concepts, value would presumably carry over to them in an unproblematic and nondiscounted fashion.

What of the cases for which the package of folk term plus associated theory can help us with more than one kind? Here we find that there is more than one kind with some of the properties that common sense deems central to a notion. Probably the boundaries of none of these kinds coincide exactly with common sense. In these situations, I claim that each of these distinct kinds is a candidate for value. We should carry over value in proportion to the closeness of the kind to the folk package.

Determining the amount of value to transfer from commonsense notions to kinds is a three-stage process. First, we must take the framework of normative and descriptive notions that make up our familiar moral theories and clarify the connections between them. This tells us exactly what place a conscious or rational being, for example, has according to widely accepted, historically resilient moral views. Next, the scientifically informed must describe the natural kinds that run through the environment. Some of these kinds turn out to be best described by using familiar ordinary-language terms. We will be especially interested in kinds corresponding with value-anchoring psychological notions.

Once we have found the best way to describe kinds, we need to decide how close they really are to appropriate commonsense descriptive concepts. To measure this distance, we ask how much and how central is the original folk theory that is preserved in our most accurate description of the natural kind. Thought experiments are called upon to separate objects falling under commonsense concepts from candidate naturalizations. If we can describe many situations that unambiguously contain the property picked out by common sense, but not the property isolated by the scientific definition, or vice versa, then we should say that the notions are relatively distant. These kinds will tend to lack some properties deemed by semantic intuition to be essential. Closer kinds will not be so frequently or obviously separable from commonsense notions.

My suggestion is that the overall distance between the commonsense notion and the scientific account determines the amount of value transferred. Kinds that are intuitively close to value-anchoring notions will inherit much value; those that are rather distant will inherit proportionately less.

I think that this approach offers significant hope to proponents of more radical accounts of environmental value. Scientific progress creates a slippage between ethical principles and co-evolved descriptive views. This allows kinds stretching throughout endangered ecosystems to compete for the moral niches once exclusively occupied by narrower kinds. In the next chapters I begin the search for plausible naturalizations fit to ground an environmental ethic.

CHAPTER 4

Recent Defenses of Biocentrism

THE VALUE OF LIFE

How can the conceptual tools developed in chapter 3 be applied to nature? The answer lies in a variant of what Gary Varner (1998) calls *biocentric individualism.* Proponents of this theory find the value of the environment in the separate and independent values of individual living things.

The argument that occupies the remainder of this book comes in two stages. First, I argue that individual living things, ranging from the rational multicellular down to the nonsentient single-celled, are intrinsically valuable. This takes up the current chapter and the immediately following one. Though such a position would clearly expand considerably on the psychological view, it is still some distance from what many would consider an adequate environmental ethic. Although many of the things that environmentalists care most about—rain forests and coral reefs, for example—are full of living things, an environmentalist concern is supposed to abstract away from component individuals to the wholes they jointly constitute. Advocates of holistic moralities deny that principles targeted at individuals could ever produce an adequate environmental ethic. Second, in chapter 6 and, in particular, chapter 7, I argue that the individualistic foundations put in place in this chapter and the next do indeed support an adequate environmental ethic. We can use individualistic language to capture what is morally distinctive about species and ecosystems.

Precursors to contemporary biocentric individualism can be found in a variety of places. Followers of the ancient Indian religion of Jainism warned that acts of violence against living beings did such damage to one's karma that they were likely to affect one's next life adversely. To bring about conformity with biocentric value, Jains encouraged *fruititarianism,* the practice of eating only such things as milk, fruit, and nuts, which could be harvested without killing the providing plant or animal.

Albert Schweitzer is the modern figure most closely associated with the view that all individual living things warrant special moral treatment. According to Schweitzer, "the great fault of all ethics hitherto has been that they believed themselves to have to deal only with the relations of man to man. . . . [A] man is ethical only when life, as such, is sacred to him, that of plants and animals as that of his fellow men."[1] Schweitzer was not entirely blind to the toll in other lives demanded by human life. Yet he urged that we kill, whether it be in the process of eating or fighting infection, only with a due sense of seriousness.[2]

The sayings and writings of the Jains and Schweitzer show it is not difficult to present life in an appealing light. It is much more difficult, however, to come up with a philosophically rigorous defense of biocentric value. Some have accused Schweitzer of offering less a fully worked out moral biocentrism than a limp biosentimentalism, and though he may have been deeply affected by the fates of termites and daisies, he was not in the business of finding watertight arguments.[3] There are serious doubts about the fitness of the concept of life to play any major role in moral theory. Compare *life* and concepts allied to it with *sentience* and the intentional notions at the heart of the psychological view. Traditional views about what agents ought to do, and what they are owed, have co-evolved with theory about what kinds of things agents are. The longevity of this relationship explains why normativity closely tracks folk psychological, or at least partially folk pyschological, notions. *Life* and the notions allied to it seem bonded to moral notions in altogether flimsier and more haphazard ways. We may be happy intoning the phrase "all life is precious," but we certainly feel in no way committed to heroic blade-of-grass rescue acts.

Only of late has biocentric individualism been systematically argued for, and my search for an adequate theory begins with an examination of the views of four biocentrist pioneers, Kenneth Goodpaster, Holmes Rolston III, Paul Taylor, and Varner. I present the argument of these pioneers in two stages. In the first stage, we are presented with a certain conception of life supposed to prepare the way for a positive moral valuation. This, the autopoietic conception, demonstrates the appropriateness of moral-sounding language in the context of life. At a minimum, it shows that talk of respecting a caterpillar is not in the same category of nonsense as advocating concern for the plans and projects of a discarded chunk of concrete bridge support. We must do more than show that we can use the words *good* and *interest* to describe living things, however. The biocentrist's next stage is to build up these goods or interests by downplaying the moral importance of the states featuring prominently in the psychological view. One style of argument is to be found particularly in the writings of Taylor (1986) and Goodpaster (1979). These philosophers argue that once sentience and rationality are placed in their proper biological and evolutionary places, we become aware that the psychological view dramatically exaggerates their importance. Appropriately diminishing these states means that they no longer obscure the value of underlying biological goals. An alternative strategy found in Goodpaster (1979) and Varner (1990, 1998) targets recent attempts to formulate the psychological view in an exclusionary way. These philosophers argue that exclusionary statements of the psychological view must pass in silence over things that patently have moral standing. Only the biocentric ethic has the resources to complete our account of the morally considerable.

The Autopoietic Account of Life

To prepare life for the moral center stage, we must first attend to definitions. In the next chapter I will make some comments concerning the overall enterprise of defining life, but for now I confine myself to providing some background for the concept motivating recent biocentric accounts.

According to Aristotle, the first systematic thinker about life, living beings can be distinguished from the nonliving in terms of their movement.[4] The nonliving falling stone follows a natural tendency to move, yet it does nothing to originate this movement. In contrast, things that are alive are sometimes governed by natural tendencies but can also self-move.

Our day-to-day theorizing about objects more or less respects the Aristotelian distinction. Commonsense physics is quite good at telling us how internally homogeneous objects of certain physical dimensions behave when acted on by visible external forces.[5] To give some rather unexciting examples, large, heavy, square objects, when pushed off a table, will, all else being equal, fall to the ground, make a loud noise, bounce just a little, and then come to rest. Lighter objects fall but tend not to make such a loud noise. Of course, we often make mistakes. For example, people are often very bad at guessing the trajectories of objects dropped from moving platforms.[6] With some knowledge about external forces, however, folk physical theories generally enable us to keep good track of medium-sized inanimate objects. Some objects seem to violate the rather general principles that make up commonsense physics in the manner that interested Aristotle. To focus on visible external forces does not seem to get us very far with them, as these objects seem to produce their own movement or activity. Among things that are recalcitrant to folk physics in this fashion will be sleeping dogs, which on being poked move in a way that is much more than a simple function of the force vector of the poking stick, or seeds, which over hours, produce shoots in response to watering. Though most people will defer to expert biological opinion, they are disposed to call such objects alive. Apparent self-movement or self-originated activity is a very important component of our commonsense concept of *life,* at least operationally.

Specifying what manner of thing could produce this self-movement has proved a tough task. Aristotle traced this capacity to originate movement back to the *psyche,* often rendered in English as "soul." This translation may be appropriate when talking about humans. However, it is clear that Aristotle intended something much broader. Though a plant's psyche may have been different from that of an animal, it was no less genuine.

Aristotle offers little guidance when it comes to saying exactly what manner of thing this self-movement promoter was. Later attempts to account for self-movement called on such blatantly nonnatural substances as immaterial vital essence or life force.

Aristotle's idea of self-originated activity is preserved in the contemporary autopoietic account of Humberto Maturana and Francisco Varela (1980). Autopoiesis refers to the ability for "self-production." An autopoietic system manages its own growth and maintains its structure, resisting the insults and assaults of the environment and, when damaged, striving to repair itself. All this occurs against a backdrop of physical flux as its constituent and surrounding matter continually changes. The scheme of mutual feedback and interconnectedness that enables diverse parts to constitute an autopoietic system is found in the individual cell, just as in the clusters of cells that jointly make up a multicellular organism. Maturana and Varela distinguish the heteropoietic from the autopoietic. Heteropoietic, or nonliving, things inherit from other beings any movement, or any productive or reproductive capacities. The products of human design are paradigms of heteropoiesis.

This autopoietic conception gives biocentrists access to a range of important moral notions. For example, it appears to legitimate Taylor's description of the individual organism as a "teleological (goal-oriented) center of life."[7] As Taylor explains, this goal orientation does not depend on the possession of any conscious or intentional states. Taylor elaborates: "[A] living being is conceived as a unified system of organized activity, the constant tendency of which is to preserve its existence by protecting and promoting its well being."[8]

He goes into a little more detail on this well-being: "Even when we consider such simple organisms as one-celled protozoa, it certainly makes perfectly good sense . . . to speak of what benefits or harms them, what environmental changes are to their advantage or disadvantage, and what physical circumstances are favorable or unfavorable to them."[9]

Holmes Rolston (1994) thinks that the autopoietic account allows us to recognize living things as unreflecting self-conservationists. As such they are worthy targets of our conserving efforts:

> A merely physical object has nothing to conserve. Though conservation of mass and energy takes place during the various events that happen to a rock . . . a rock conserves no identity. It changes without conservation goals. . . . Biological organisms, by contrast, conserve an identity—a metabolism maintains itself and an anatomy over time. Organisms have a life, as physical objects do not.[10]

This is just to sample the purported linkages between the autopoietic conception and moral notions.

So terms such as *good, goal, interest,* and *conserve* prominent in moral contexts can also be used in talk about living things. But is there any assurance that the autopoietic conception isn't just encouraging us to conflate two distinct usages of these terms, thereby promising moral conclusions on the sly?

The theoretical apparatus described in chapter 3 might indicate if this is the right path. Perhaps it leads to plausible naturalizations of moral notions. The main problem with this proposal is that the autopoietic conception transfers moral-sounding language without significant normative-descriptive linkages. We are used to thinking of morally significant interests or goods as being accompanied by plans, projects, and potential experiences. The autopoietic conception does not claim to find in living things intentional states or plausible naturalizations of such states. And this makes somewhat hit-or-miss any transfer of value.

A Life Ethic or an Everything Ethic?

Evidence for the flimsiness of the connections between autopoietic goods and morality comes from Janna Thompson (1990). She argues that the account of interests emerging from the autopoietic account of life is dangerously liberal—leading to moral nonsense. In conforming with the common-sense alignment of morality with psychology, we get the benefit of millennia of folk ethical experimentation and consideration of problem cases. Though we may have a few vague cases to deal with, there is a good chance we can make this boundary fairly principled. This seems not to be the case with the life ethic. Rocks, rivers, and molecules are not alive, but they seem capable of being understood as having goods in the biocentrist's worldview. These objects have certain structures enabling them to resist some threats from outside. The rock will not shatter when subjected to just any blow; the river washes away the mud slide that would otherwise divert it; and so on. Think of the implications for the moral status of the occupants of ALH84001. The life ethicist would want to say that discovering that the carbonate globules were once alive makes a big difference to the historical moral status of the meteorite. Yet regardless of whether a given globule was ever a Martian microbe, it will still tend to maintain its structural integrity, and certain interactions with its environment will act against this integrity, or "harm" it.

Perhaps we should not be overly concerned about this threatened ethical explosion. According to Schweitzer, the moral person not only "tears no leaf from its tree, breaks off no flower, and is careful not to crush any insect as he walks" but also "shatters no ice crystal that sparkles in the sun."[11] Schweitzer's readiness to lump the goods of ice crystals in with those of plants is ill-advised. We need to draw and defend the moral significance of a boundary between living and nonliving things to prevent our life ethic from degenerating into an everything ethic. If absolutely everything is to be special, then gone will be any moral incentive to look out for nature. What grounds then would permit us to limit inherent worth, or intrinsic value, to biological objects, thereby marking off the microbe's good from the nonbiological carbonate formation's and ice crystal's goods?

Varner (1990, 1998) has an argument that might make this restriction of value principled. He counters the apparent arbitrariness of assigning interests to nonsentient beings by appealing to a biologized notion of interests. In his view, a being possesses interests if it has subsystems with biological functions. Varner accepts the etiological account of such functions discussed in chapter 3. To take his example, a cow's mammary glands have the function of producing milk for calves rather than having some other function, such as the boosting of dairy farmers' profits, because the glands have been selected to permit the nourishment of calves.[12] Varner's aim is a determinate story about the interests of the nonsentient constructed on the foundation of this functional determinacy. In carrying out its function, a biological device contributes to the interests of the organism that houses it. By the same token, activity that impedes the biofunctions of an individual's subsystems acts against the interests of that individual.

This account allows us to get around one set of Thompson's problem cases. Rocks, unlike nonsentient living individuals, clearly do not possess interests understood in Varner's way, as they do not have parts with selection histories. It is simply not the case that a particular groove on a rock exists now because of the efforts of similar grooves on earlier copies of the rock.

Varner's move may help with another objection raised by Thompson. Parts of organisms, such as kidneys, lungs, and hearts, have their own goods that the environment can damage. Does every part of any living thing, therefore, have its own separate or intrinsic value? The obvious response will be that these interests exist only in the context of the organism. For example, the selectionally determined function of the kidney is to purify blood *for its possessor.* This answer does not impress Thompson, however. She draws an analogy between the relationship the good of a kidney bears to its possessor's good on the one hand, and, on the other hand, the relationship the good of a wood-boring insect bears to the good of the tree in which it is found. The good of the kidney ". . . can be determined independently to the same extent that the good of a wood-boring insect can be determined independently of the good of the tree it feeds on. . . . Kidneys, like insects . . . need certain kinds of nourishment; they are healthy under some conditions and are caused harm by others. These conditions can be specified without mentioning the organism in which the organs reside."[13] Attention to the historical dependencies at the heart of the selectional theory of function allows us to see a key difference in the ways kidneys and wood-borers relate to the organisms that house them. In both cases we can, as Thompson insists, give a statement of good or function that makes no reference to the good of any surrounding organism.

However, though the selectional theory permits this narrower statement of my kidney's good independent of my good, it also validates a specification in terms of my good. In saying that kidneys have been selected to purify blood, we make no direct reference to the good of any individual organism. We can go on to ask another question, however. Why does blood purification occur, or what is *its* selected function? An expanded selectional account will, in the end, refer to the enhanced survival and reproductive prospects of human bearers, making the link between kidneys, blood purification, and enhanced fitness. No such expanded account is possible in response to a question about the relationship between the wood-borer and the tree. The contemporary wood-borer's existence does *not* depend on past wood-borers' having produced some benefit for past trees. Indeed, wood-borers are unlikely to have boosted the fitness of the trees that they infested. It follows that on no valid expansion of the selectional account is it the function of any current wood-borer to contribute to the good of any tree.

There are two potential difficulties for those who would supplement an autopoietic account of interests with the etiological theory of function. The first, more practical concern is that the two components of our new composite theory seem to pull in different directions. In chapter 7 I will discuss cases of biological altruism as I build out from the goods of individuals to an ethic of species and ecosystems. Such cases are bad news for the proposed autopoietic/biological functional amalgam. The bee that dies in stinging a hive invader does something that it was selected to do— therefore, on the etiological account, something in its interests. Yet it is hard to see how the autopoietic account could recognize the resulting biological breakdown as in the bee's interests.

This schizophrenia about interests that emerges from our merging of Taylor's and Varner's theories leads to a more theoretical worry. A selectional theory of function might be a dubious fallback if a widely supported view of natural selection turns out to be true.

I have argued that the selectional theory of function can establish a teleological link between an organism and its parts. I suggested that the ultimate goal of my kidney is my good. I took the orthodox Darwinian view and assumed that these ultimate goods would be those of individual organisms. This position has come under fire over the past thirty years. According to G. C. Williams, Richard Dawkins, and others, natural selection, the principle organizing force in the biological world, operates at the level of the gene.[14] For genic selectionists, evolution by natural selection is most fundamentally a story about directed changes over time to the constitution of the global gene pool; the function of each chunk of phenotype is ultimately the enhancement of the replication prospects of underlying DNA. We should expect this to have unwelcome consequences for the position discussed above. If genes are targets of selection, then the only function or purpose in nature is their preservation. An ethic grounded in the maintenance of biological goods will place genes above individual organisms, and it will be a Siberian tiger's DNA, rather than the tiger itself, or its species, that is worthy of respect.

The prospect of a moral gene-centrism is a serious threat to any environmental ethic. As the account I favor has a substantial natural-selective component, this issue will need to be addressed. I do this in chapter 6. At this point, I will say that tools exist to help describe what is in the interest of a nonsentient organism, though these interests may fit awkwardly at the heart of an ethic.

Sentience and Rationality Placed in Their Proper Biological Places

One way to build up the supposed moral goods of the bacterium or the fungus is to morally "rightsize" sentience and rationality. This is the strategy of Goodpaster (1979) and Taylor (1986).

Goodpaster (1979) has this to say about the place of sentience in the biological world: "Biologically, it appears that sentience is an adaptive characteristic of living organisms that provides them with a better capacity to anticipate, and so avoid, threats to life. This at least suggests, though of course it does not prove, that the capacities to suffer and enjoy are ancillary to something more important rather than tickets to considerability in their own right."[15] The human capacity for pain and pleasure did not arise as an end in itself. We get biological priorities right if we say that it is of secondary importance to one or other or some combination of autopoietic and naturally selected ends.

We can parallel this type of biocentric argument with Charles Darwin's response to utilitarianism.[16] Happiness should not be morality's end, said Darwin, as to think so was to invert the proper evolutionary relationship between survival and reproduction on the one hand and happiness on the other. Emotional states such as happiness and suffering were of value only insofar as they promoted the survival prospects of community members.[17] In parallel with Darwin's charge that Mill gets the wrong way around the relationship between happiness and survival and reproduction, biocentrists accuse various exponents of the psychological view of inverting the proper biofunctional relationship between the goods illuminated by the biocentric approach and sentience or rationality.

Biocentrists are not of one mind on the issue of what strength of ethical conclusion to draw from this observation. The less thoroughgoing option is to insist that while biological goods must be accepted as the universal requirement for value, the presence of rationality or sentience adds value. Goodpaster (1978) leaves this option open by distinguishing between moral considerability and moral significance. Though a human and an ant cannot be distinguished in terms of their moral considerability, we are permitted to say that being folk psychological will make the human more morally significant.[18]

This compromise position sits uneasily with the general tenor of Goodpaster's biological argument. Compare the reasoning with Taylor's more thoroughgoing approach. He advocates a biocentric egalitarianism, rejecting any moral specialness for sentience-assisted pursuit of fundamental biological ends. Had the environment posed different challenges to our distant ancestors, then armored plates or acute night vision might have evolved in the place of rationality. These are all means to identical biological ends. Taylor complains of any attempt to reserve a special place for mental traits:

> It is not difficult to recognize a begging of the question. Humans are claiming human superiority from a strictly human point of view, that is, from a point of view in which the good of humans is taken as the standard of judgement. All we need to do is look at the capacities of nonhuman animals (or plants for that matter) from the judgement of *their* good to find a contrary judgement of superiority. The speed of the cheetah, for example, is a sign of its superiority to humans when considered from the standpoint of the good of its species.[19]

The members of most species are better than the members of other species at something, and according to Taylor, we are wrong to think that one set of properties that happen to be possessed to a high degree by humans confer either base-level moral value or a value bonus.

If one is determined to tie the moral purchasing power of sentience to its place in nature, then Taylor's conclusion would seem easier to arrive at than Goodpaster's. One could even imagine an argument starting from the fundamental importance of biological ends and moving to the *reduced* importance of sentience or rationality. Capacities such as flight and echolocation are kept on a relatively tight leash by underlying biological needs, and only on rare occasions will they act against these needs. In contrast, humans frequently use their rationality in ways that compromise their biological needs. It guides them toward vasectomies and bungy jumps, for example.

Later I will explore the debate between monistic egalitarian and pluralistic hierarchical formulations of the life ethic. For now, I must question the biocentrist's claim that this biological diminishment converts into moral diminishment. The dominant reasons for believing in the psychological view are certainly not evolutionary or biological. Opponents of evolutionary ethics will argue that claims about natural selection are powerless to undermine the psychological view.

Biology is not the only empirical discipline with its own scheme of normative interrelationships. Exercise physiologists and economists, for example, postulate their own ideal sets of circumstances and have their own ideas about how the psychological states revered by commonsense morality can help bring about these states. The economist may hope that the individual uses rationality to promote market efficiency; the exercise physiologist will want mind to direct body so that oxygen intake is optimized and so that arms and legs are moved in coordinated fashion. We can accept the theoretical importance of these goal states and claims about how they are best brought about while rejecting the calls of the economist to assign greater moral significance to markets than to individuals or of the exercise physiologist to value individuals' particular plans and projects less than the body beautiful and well exercised.

The same response can be made to the biocentrist. It may be the case that rationality and sentience are *biologically* less fundamental than the seeking of survival and reproduction; they are biologically "for" the preservation of the organism and the production of offspring. The ethicist can still say that biological subsystems more intimately connected with survival and reproduction are *morally* "for" the seeking of happiness or the fulfillment of preferences. We need better reasons to take biological interests or goals seriously.

The Incompleteness of the Psychological View

According to the arguments I will now discuss, the psychological view is *incomplete*. It neglects things that must morally count for something. Only the addition of biocentric value can remedy this deficiency.

Goodpaster's (1978) version of the incompleteness argument targets the recent exclusionary statements of the psychological view due to Singer, Joel Feinberg, and G. J. Warnock. According to these philosophers, beyond the rational or the sentient there is nothing to account for. Goodpaster responds that their positions fail to acknowledge "in nonsentient living beings the presence of independent needs, capacities for benefit and harm, etc."[20]

For example, Feinberg (1974) frames his restrictive account of morality in terms of rights. He claims that the possession of interests is a necessary condition for rights, continuing that interests in turn presuppose desires. Feinberg argues that without recourse to desires, we can make no sense of a plant's having an interest and that this must mean that the plant can have no moral rights.

Goodpaster appeals to the biocentrist's story about biological interests to challenge this assertion: "There is no absurdity in imagining the representation of the needs of a tree for sun and water in the face of a proposal to cut it down or pave its immediate radius for a parking lot. . . . In the face of their obvious tendencies to maintain and heal themselves, it is very difficult to reject the idea of interest on the part of trees (and plants generally) in remaining alive."[21] This argument of Goodpaster's leaves us with our original conundrum. We know that we can use the term *interest* when talking about trees. What reasons do we have for thinking that these are distinctively moral interests? We have as much reason to think that every interest carries moral weight as we do to think that money can be withdrawn from anything we call a bank.

We need further argument before we feel forced to give arboreal interests moral weight. Varner (1990, 1998) seeks to do this, adding a novel twist to the incompleteness argument. Rather than reaching immediately for the moral values of stick insects and louse worts, he takes as his starting point objects whose value is not disputed by the psychological view: persons. According to Varner, to adequately characterize the value of persons, we will have to appeal to capacities that stand in no direct relationship with sentience or rationality. Such an acknowledgment leads directly to biocentrism. These capacities permeate the nonsentient living world and, therefore, are capable of underwriting its nonanthropocentric value.

Varner begins by describing what he calls the dominant version of the psychological view (or, as he puts it, the Mental State Theory). This version of the theory equates an individual's interests with her or his desires, each desire corresponding with an interest. Though simple versions of the theory will look to an individual's actual desires, more sophisticated variants ask what she or he would desire were relevant information available for consideration and were she or he to be impartial across different phases of life. This hypothetical version enables us to discount many actual desires. A desire to eat ten tons of chocolate cake for breakfast is presumably not one that would survive rational scrutiny.

Varner then seeks to show that this best exposition of the psychological view has holes. His chief example is Maude, "an unusually intelligent and generally farsighted young adult who has a strong desire to smoke."[22] Varner supposes that this desire to smoke survives our apprising Maude of all the relevant information about the dangers to health of smoking. He continues:

> On the mental state theory, no sense can be made of the claim that Maude's smoking is bad for her—that is, contrary to her interests. This is because she does not now desire to stop smoking, and on the mental state theory of harm this implies that continuing to smoke is only bad for her only if her enlightened preference would be to stop smoking. But, by hypothesis, Maude is both adequately informed and impartial across all phases of her life. Therefore, her actual preference is her enlightened preference, and therefore on the mental state theory of harm, Maude's smoking is in no way bad for her.[23]

Yet Varner thinks this is a crazy response. There is an unmistakable sense in which Maude's smoking is bad for her, impeding her interests.

He gives a further example of nineteenth-century seafarers and their need for ascorbic acid to combat scurvy. These people knew nothing of the prophylactic properties of vitamin C. Unless one were to make very unrealistic assumptions, making available desires that went far beyond the ken of the contemporary science, then we should not ascribe to the sailors even a hypothetical desire for ascorbic acid. Yet it would seem clearly in their interests for them to have consumed vitamin C.

The etiological theory of function enables us to finish the job of describing individuals' interests started by the desire theorist. Smoking acts against certain of Maude's interests because it stands to impede the proper functioning of the various subsystems of her body—her heart and her lungs, for example. By the same token, the failure to consume vitamin C sets back interests by leading to the anemia, spongy gums, and tooth loss associated with scurvy.

Suppose we concede that any account adverting exclusively to mental states fails to capture all of a person's morally significant interests. Now, to be consistent, once we recognize the moral importance of these interests in human beings, we must also acknowledge their importance in nonhumans. If so, interfering with the well-functioning subsystems of plants and insects ought also to harm morally significant interests.

Varner's argument is a significant advance over Goodpaster's. It seems we can no longer lump plants together with economies, air-conditioning systems, and ice crystals in the category of things that have interests either of no moral account or of exclusively anthropocentric import. An adequate theory of anthropocentric value seems to contain within it the seeds of biocentrism.

Varner's conclusion depends crucially on whether the story about Maude and the mariner-scurvy case really rules out exclusive reliance on the desire theory of interests. I suspect that the desire theory is better resourced than Varner supposes.

So how can the desire theory say that the satisfaction of Maude's desire to smoke, resistant to rational revision or elimination as it is, is not in her interest? Varner's requirement of impartiality across different phases of life provides one opening.

We can understand this impartiality in two ways. There is an *internalist* reading. Here, different phases of life will have equal impact on Maude's reckonings; information about potential harms and benefits associated with different life stages will be represented with equal weight. This is the reading that Varner supposes—Maude is such a determined smoker that her desire can survive vivid representations in her thinking about possible future ill-health. Contrast this with an *externalist* way of understanding impartiality across different phases of life. By this reading, we place on the table a person's considered desires, both present and future. We note that the satisfaction of a given desire differentially affects other desires. It may bring some desires closer to satisfaction at the same time as tending to set back others. We will not be content with the agent's estimate of which interests to sacrifice in case of a conflict. Rather, our goal is the optimal combination of desire satisfactions available to an agent given her or his overall actual and potential preference structure. While we will still say that each considered desire generates an interest, we calculate an individual's overall interests by emphasizing those interests that in fact bring about as favorable as possible a balance of desires satisfied over desires frustrated. Often, one of a person's interests will conflict with his or her all-things-considered interests, regardless of the attitude adopted toward it.

Returning to the story of smoking, the externalist says that a desire to smoke goes against Maude's interests, all things considered, if satisfaction of it has the propensity to frustrate other desires central to her life plan. This might be the case if Maude also desires to live a long and healthy life, to go on bridge cruises as a seventy-year-old, and so on.

Note the overlap between biological interests and the externalist desire account. If the satisfaction of a given desire goes against what Varner understands as Maude's biological interests, it will likely also make other central life-plan-related desires vulnerable. Most people will best pursue their future goals if they are fit and well. However, the biological interests account and externalist desire theory will give different answers to some questions about interests. Suppose, implausibly, that Maude's central desires include the plan to leave a corpse suitable to scientific research into smoking-related desires and the hope that she not see the year 2010. Suppose also that she has no significant desires that conflict with these. In this case, we might say that Maude's smoking does not tend to set back any interests visible to the externalist desire theorist while thwarting biological interests. I suggest we should also say that the desire to smoke will not set back morally relevant interests.

What should we say about Varner's scurvy-menaced sailors? Again, we gather together considered desires, noting that, for reasons entirely mysterious to the sailors and ship's doctor, those desire mixes that contain the preference to go on shorter expeditions, to make frequent port calls, or to sail on citrus-laden ships contain a more favorable balance of desires satisfied over desires frustrated. So it is in the interests of sailors to consume vitamin C.

An externalist interpretation of the desire theory allows us to deal with a further complication of Varner's problem cases. Varner wonders whether it is in the interests of his cat Nanci to go outdoors. Though Varner is happy with, and indeed defends the idea of feline desires, he is rightly skeptical about feline *considered* desires. Cats know nothing nor are capable of being instructed about the risks of fleas and feline leukemia virus (FeLV). Though only crass insensitivity to feline cognitive limitations will lead us to speculate about what the cat would think upon learning of the horrors of FeLV, an externalist reading is available. Were the cat to fulfill its desire to go outside, how would this affect other present and future cat desires? If getting FeLV is part of the price of prowling the garden, then we should expect numerous and significant cat desires to be frustrated later. It is, therefore, not in the overall interests of the cat to be let outside.

AN ETHIC TO LIVE BY?

In the preceding section, I have investigated the argumentative support
for biocentric value. I have argued that biocentrists have not yet given
enough for us to abandon the received moral wisdom encapsulated by the
psychological view. I will shift now from the discussion of moral inputs—
the reasons we are given to believe in certain values—to the examination
of moral outputs—whether a given value is really capable of informing
behavior. My discussion begins with the following obvious riposte to the
biocentrist.[24] Surviving means killing to eat and allowing our immune
systems to slaughter millions of invading bacteria. It is, therefore, pragmatically impossible for us to respect the values of all individual living
things.

This fact of life presents the biocentrist with an awkward choice. One
possibility is that the value of the nonsentient alive is so burdensome as to
make biocentrism an ethic for God but certainly not for humans; God
presumably has the dual advantages of omnipotence and of not having to
live cheek by jowl with his or her creations. Alternatively, we may seek to
lighten the biocentric load. Attempts to make life-value more humanly
tolerable threaten to make it, though perhaps a theoretical reality, almost
certainly also a practical irrelevancy. We may talk in biocentric terms inside
the classroom, but we will be well advised to be guided by some expression
of the psychological view when dealing with real-world problems.

A distinction will help us recognize the wider ethical options available
to biocentrists. Biocentrists can subscribe to either a *monistic* or a *pluralistic* approach to moral considerability. A monist such as Taylor thinks that
once we have taken into account an organism's life-related value, we have
all we need to make comparisons with other objects. Pluralists differ in
supplementing biocentric value with other brands of value. Goodpaster,
Rolston, and Varner all believe that being sentient or rational contributes a
distinct value to a being's moral makeup.

Taylor's monism leads directly to a moral egalitarianism about living
things. As we saw earlier in this chapter, the rejection of any value hierarchy in nature is essential to his position. Taylor sees no argument for the
increased importance of humans, whether it be by way of claims about
sentience or rationality, or about any other property, that does not depend
on some species-specific bias. All living beings are equally inherently worthy: "To accept [the biocentric outlook] and view the world in its terms is
to commit oneself to the principle of species-impartiality. No bias in favor

of some over others is acceptable. The impartiality applies to the human species just as it does to nonhuman species."[25]

So, all living beings have goods. There are no grounds for thinking that the good of one being should count for more than the good of another, and as a result, we must consider all living beings to have equal moral significance.

To make his ethic manageable, Taylor lodges biocentric value in a patchwork of nonconsequentialist principles. It seems clear why he should adopt such an approach. If, as the consequentialist does, we reject the moral distinction between acts and omissions, or that between the intended and foreseeable consequences of our choices, the demands of the merely alive threaten to overwhelm us. In setting out to walk across a park I do not intend the deaths of any living beings, yet it is the case both that there are courses of action available to me that would better promote the interests of all living beings, impersonally considered, and that I impose on park-living insects the easily foreseen risk of being trodden underfoot.

Despite the best of Taylor's efforts, he does not avoid intolerable conclusions. We can begin with his principle of self-defense. This allows that we can kill both sentient and nonsentient beings when doing so is essential to the continuation of our own lives.

> The principle of self-defence states that it is permissible for moral agents to protect themselves against dangerous or harmful organisms by destroying them. . . . The principle does not allow the use of just any means of self-protection, but only those means that will do the least possible harm to the organisms consistent with the purpose of preserving the existence and functioning of moral agents.[26]

So we can allow our immune systems to continue annihilating salmonella bacteria, and we can kill to eat. Taylor's reference to moral agents may seem to be placing humans above other organisms. He denies this. His principle of self-defense is supposed to be species-blind: "Despite what might at first appear to be a bias in favor of humans over other species, the principle of self-defence is actually consistent with the requirement of species-impartiality. It does not allow moral agents to further the interests of any organism because it belongs to one species rather than another. In particular humans are not given an advantage simply on the basis of their humanity."[27] Taylor's requirement of species-impartiality is given teeth by a rule of noninterference:

> We are . . . required to respect [organisms'] wild freedom by letting them
> alone. In this way we allow them, as it were, to fulfill their own destinies.
> Of course some of them will lose out in their struggle with natural com-
> petitors and others will suffer harm from natural causes. . . . If we accept
> the biocentric outlook and have genuine respect for nature . . . we remain
> strictly neutral between predator and prey, parasite and host, the disease-
> causing and the diseased. To take sides in such struggles, to think of them
> in moral terms as cases of the maltreatment of innocent victims by evil
> animals and nasty plants, is to abandon the attitude of respect for all wild
> living things.[28]

The principle of self-defense constrained by the requirement of species-
impartiality and rules of noninterference leads to some problems. What
should our attitude as third parties be to conflicts between humans and
other living beings? The bacterium *vibrio cholerae* causes cholera. Many
would claim that intervention on behalf of cholera-stricken humans in
distant communities is morally worthy. Yet for the biocentrist, we have
morally valuable humans on the one hand and equally morally valuable,
but far more numerous, *vibrio cholerae* bacteria on the other. It is morally
permissible for infected humans to cure themselves, but in assisting them
we fail to act in a way that is impartial between species.

To arrive at what we are strongly disposed to judge the right response
to the above situation, we need to find grounds for giving humans more
moral weight than bacteria. As *vibrio cholerae,* mollusks, giraffes, and
humans are all equally alive, it is hard to see how monistic biocentrism
could provide such grounds.[29]

The monist suffers grievously from the failure to adequately morally
motivate life. Though startling and demanding conclusions are certainly
not unheard of in moral theorizing, when we find them we do require a
high standard of argument. If we had made a better initial case for the
moral significance of life, we would be more likely to be guided by species-
impartial principles than to treat them as a reductio

Is Value Pluralism the Answer?

An obvious response to these difficulties would be to reject Taylor's monistic biocentrism and supplement life-associated worth with more familiar
brands of value. Such an account will allow the greater intrinsic value of
the rational and conscious to outweigh the value of the merely alive, and

so avoid the absurd consequences of Taylor's theory while still assigning value to all living things.

Rolston (1988) is one who pushes for this type of pluralism. He thinks that though all living things are intrinsically valuable, human mental attributes give us "the highest per capita intrinsic value of any life form supported by the [eco]system."[30] As we saw, Goodpaster (1978) distinguishes between moral considerability and moral significance to allow that though both humans and slugs are equally morally considerable, the human is more morally significant.

It is easy to see the appeal of pluralistic approaches in ethics. Familiar monistic options, such as hedonistic utilitarianism, provide far too crude a measure of a human life. There has to be more to a worthwhile human existence than the accumulation of units of pleasure. By introducing independent intuitively attractive values (for example, wisdom, truth, beauty), we arrive at an account that comports more readily with strongly held convictions about the good life. There is a penalty to pay, however. The greater the diversity of values, the more frequent the apparently irreconcilable clash of incommensurable values.[31] This is particularly worrisome for the biocentrist.

An amalgam made up of both biocentric value and familiar sentience or rationality-based value requires some mechanism for systematically comparing these lower and higher goods. Such a need is more pressing when our main concern is to establish the credentials of an unfamiliar value as part of the pluralistic mix. Pluralism about human ends seems plausible because each of the supposedly conflicting and incommensurable human goals has a relatively secure place in our affections. The same is not true of life-value. Without some principled means for ranking biocentric value alongside human-centered value, even the most fleeting and trivial human desire may end up deserving more attention than the life of a nonsentient being. Humans are finite beings whose actions are capable of subserving only a finite number of goals. If the choice is between courses of action that affect only nonsentient beings, our value pluralist may urge us to decide in a nature-respecting way. How often will this be the case? Advancing technology opens an increasing array of environment-consuming goals to ever-growing numbers of humans. Failure to adequately calibrate human and nonhuman value may mean that there will almost always be more worthy recipients of moral concern than nonsentient nature. Familiar human-centered value threatens, for all practical purposes at least, to drive this new biocentric value out of existence.

So, we lack a story about how the interests of the merely alive are to feed into the overall moral calculus together with the weightier, much more difficult to ignore interests of the sentient and the rational. If we cannot find such a story, life-value seems likely to be perennially shouted down. The problem is that the apparatus of pluralism cuts the biocentrist off from the theoretical resources necessary to sustain the sought-after hard-hitting conclusions. The pluralist holds that there are at least two values that morally correct behavior must respect, biocentric value and human-centered value. What can we say about the relationship between these values? Beyond knowing that human value is higher than and trumps biocentric value, we have no means of translating one into the other. If this is all we can say about the relationship, then we can also claim to be respecting nature's value in giving preference to even the most unimportant human interest.

Varner's Pluralism

I come now to a pluralistic account that promises both to leave adequate room for human projects and to give nonsentient beings genuine moral weight. Varner (1998) describes a three-tier hierarchy of beings with interests. Most important for Varner are those individuals with ground projects. Following Bernard Williams, he understands a ground project as "a nexus of projects . . . which are closely related to [one's] existence and which to a significant degree give a meaning to [one's] life."[32] This gives Varner the nonspecies-ist means to defend the moral preeminence of most humans. Almost all humans, perhaps some of the great apes, but certainly no other nonhumans on this planet, have, or will have, ground projects. Next on Varner's list are beings with desires insufficiently sophisticated to combine to form ground projects. He devotes a chapter to the defense of mammalian desire, also considering it likely that birds and octopuses have desires and possibly that reptiles do too. Organisms not sufficiently, representationally, and behaviorally sophisticated to possess desires occupy the lowest tier of Varner's value hierarchy.

It is one thing to rank organisms within a hierarchy but quite another to say how we should choose between organisms whose most morally significant interests are found at different levels on this hierarchy. Varner discusses and rejects a broadly utilitarian reading according to which the prospect of satisfying sufficiently many relatively low-level interests justifies us in frustrating a higher-level interest. Instead, he places the pursuit of any one ground project ahead of the satisfaction of any number of desires of organisms insufficiently complex for ground projects and nonintentional biological interests. According to Varner, "it is better to eat nonhuman organisms and thereby doom all of their interests than to doom one's ground project."[33]

Though the satisfaction of higher-level interests will always trump the satisfaction of lower-level ones, Varner offers the following principle that promises to give the satisfaction of biological interests some pull: "Other things being equal, it is better to satisfy ground projects than require, as a condition of their satisfaction, the dooming of fewer interests of others."[34] Varner applies this principle to a story of Mary Midgley's about an individual who desires to torch a forest to see how it burns. This person ought to abandon such desires in favor of those that do not doom so many biological interests. Varner supposes that he can find other ways of entertaining himself or "executing a grand and dramatic kind of performance art in nature" that are kinder to trees.[35]

There is an interesting practical upshot of Varner's conjoining of the claim that ground projects come first with the requirement that we prefer ground projects associated with a favorable balance of lower-level interests promoted over lower-level interests doomed. We now have a prima facie reason to prefer a vegetarian diet over a nonvegetarian one. The killing and eating of living things is undoubtedly essential to the pursuit of human ground projects. This said, a diet that dooms only biological interests is better than one that frustrates the morally more significant desires of sheep and cattle.

Earlier, I described a problem for value pluralisms that propose a hierarchy of values. Without some mediating principle, the interests of those occupying higher strata threaten to monopolize moral concern. Varner's requirement that we opt for desires that doom fewer lower-level interests does seem to help us out of this problem, promising to protect human moral specialness in a way that looks out for the value of the nonsentient alive.

But Varner's principle does not do enough. Like other pluralists, he succeeds in giving sufficient weight to human interests only by scaling back biocentric value to a practical irrelevancy.

In chapter 5 I will make use of a distinction between the self-directed and other-directed preferences of nonhuman organisms. Self-directed preferences may bring about the frustration or satisfaction of other organisms' preferences along the way to their satisfaction; however, these other organisms' preferences do not feature in their contents. In contrast, other-directed preferences cannot be satisfied without also satisfying the preferences of some other organism. The interests of other organisms are either built into their contents or directly implied by these contents.

Later, I will rest a great deal on the other-directed states of nonhumans. For now, human desires and ground projects are my concern. It is an uncontroversial observation that humans are not islands and our interests cannot be understood in isolation from other human interests. A typical human ground project has significant other-directed components. For example, the central desires of many parents are focused on the pursuits of their children. People join in cooperative ventures—businesses, academic departments—that further entwine ground projects. The desire that my department get a good writeup by the review panel may not specifically mention the desires of other people. It is, however, quite directly connected to the projects of others.

I think these unexciting claims spell bad news for Varner's undertaking to make biocentric value count. Suppose I am in the position of having to settle upon some desires to become central to my ground projects. I find I have some time on my hands and an expectation that I will have free time over the next several years. I can adopt a hobby or a cause. Among my options are bridge-playing, support for UNICEF, or Greenpeace membership.

How am I to make my choice? One question concerns the extent to which I should be guided by moral considerations in setting up my ground projects. If I am to be guided by them, I would assess each of bridge-playing, Greenpeace membership, or UNICEF support in terms of its contribution to appropriate moral goals. Varner offers some guidance on how to determine what is an appropriate moral goal. I must ask how my choice will affect the interests of other living beings. I am to be guided in cases of conflict by the rule that higher-level interests are to trump lower-level ones, the order being ground projects, then simple desires, and finally biological needs.

Here is where the overlapping web of ground projects spells bad news for Varner's undertaking to allow biological interests to sway moral deliberations. If we accept Varner's priority principle, we should look out for the interests, however trivial, of all beings with ground projects before we look to nonsentient nature. The desires of friends (human ones), relatives, and the famine stricken for art deco homes, parts for Playstation game consoles, and food should all come before the needs of keas and horseshoe crabs. Given that there are so many human interests, it seems unlikely that moral considerations could guide us to a project centered around biocentric value.

Let's return to the desire to torch native forests, thereby producing a grand spectacle. We should begin by conceding that such a spectacle comes at a substantial cost in terms of biological interests. What of the other interests that might be contributed to? Many people may desire to witness or read about such an unusual natural performance art. We can even suppose that art critics and commentators will see the burning of the forest as contributing to a ground project. To bring a new and daring style of artwork to the world's attention might be a life's ambition. Now, all things considered, it might have been better that art critics and arsonists had more nature-friendly preferences. However, even though they have preferences that take such a heavy toll on the nonsentient alive, we must place them, in our moral reckonings, above the needs of kauri and forest-litter dwellers. Certainly, their needs cannot be modified by the would-be arsonist in the way he or she might alter his or her own desires.

The overlapping nature of human ground projects and desires will make such a case the rule rather than the exception. There would always seem to be human preferences to place ahead of the needs of the environment.

Now, believing that moral considerations should exclusively, or substantially, shape ground projects is perhaps to fall into the trap of moralism. Any life worth living must have nonmoral commitments. If we accept this criticism, we should allow the committed environmentalist to use space quarantined from morality to pursue the interests of nonsentient interests. We would have to acknowledge, however, that to do so would be to follow a morally suboptimal path in the formation of our commitments. Certainly, it would mean that the nature-phobe is under no compunction to take account of biocentric value.

Varner has not given biocentric value secure enough a place in ethical theory. He himself does see the need to compensate for the motivational deficiencies of biocentric value, and in the final chapter of his book he bolsters the environmentalist case by appealing to morally special human ground projects, arguing in classic anthropocentric fashion that current and future human ground projects require an intact nature.

We appear to have trodden a biocentric path back to a moral anthropocentrism about nature.

CHAPTER 5

ॐ

A Morally Specialized Account of Life

COMMONSENSE AND CUSTOMIZED
ACCOUNTS OF LIFE

In the last chapter I have surveyed recent biocentric ethics, identifying two types of problem. First, substantial reasons are needed to overcome traditional morality's reluctance to go beyond various expressions of the psychological view. Thus far biocentrists have failed to give the goods or interests of the nonsentient alive sufficient moral oomph to accomplish this task. When faced with a seeming problem case, the traditional moralist could argue either that, appearances notwithstanding, the psychological view is sufficiently elastic to cover such cases, or that the candidate goods or goals, though biologically well-founded, count for nothing morally. The second problem takes the form of a dilemma. If biocentrists adopt Taylor's monistic egalitarian approach, they end up with an ethic that gives inhumanly much weight to biocentric value. With Varner's, Goodpaster's, or Rolston's approaches, an advocate of the psychological view could confidently concede the genuineness of biocentric value but argue that it is for all practical purposes irrelevant. Both the biocentrist pluralist and the advocate of the psychological view will end up giving the same practical advice to someone faced with any choice between nature-respecting and human-respecting courses of action. Some kind of middle path between the rigors of monism and the permissiveness of pluralism seems to be the answer, but how are we to make such a path principled?

The first topic of this chapter is the enterprise of defining life. The pluralism I defend makes room for a range of distinct concepts of life, each one tailored to a specific theoretical end. Among these is a morally specialized conception. I propose to make good on biocentrism's motivational deficiencies by avoiding a starting point conceptually alien to the psychological view. The representational account of life I favor is composed of plausible naturalizations of value-laden folk psychological notions. In this way the values of the nonsentient alive become visible to the partisan of the psychological view.

The representational conception of life also allows a solution to the practical problem identified in the second half of chapter 4. I claim that this notion not only enables us to spread value very broadly throughout nature, but also contains a mechanism for resolving conflicts generated by the inevitable competing claims. It establishes a continuity leading from believers and desirers, the most straightforward deservers of moral consideration, through to simple living things. We need to do more than find a natural continuity between those objects to which we are disposed intuitively to assign moral value and those to which we are not, in order to draw conclusions about value assignments. In placing an organism on this continuum, we demonstrate an important relationship between it and familiar value-endowing descriptive notions. Locating an organism on this continuum also determines the amount of value we should assign to it, generating a single moral currency with which to decide the competing claims of different ecosystem members.

In the last chapter I have quickly traced the history of efforts to define life from Aristotle to Maturana and Varela. I want now to take a step backward to reflect upon our definitional ambitions. I will argue that it is a mistake to aim at an all-purpose definition of life, as there is no reason to expect any single modern account can fulfill all of the theoretical interests that arise in connection with life. Aristotle could not have anticipated the diverse explanatory and predictive needs of modern molecular, population, and developmental biology. We should, therefore, look to meet the varying needs of science without paying heed to any philosophical demand for definitional unity. Each different plausible naturalization of life should build upon an important strand taken from common sense so as to establish clear links to central notions within a target area.

To illustrate how abandoning definitional monism about life is likely to facilitate progress, I will outline two scientifically useful plausible naturalizations of the concept.[1] First, we have a metabolic notion of life. This account requires the possession of a metabolism, machinery whose job it is to translate material originating from outside the individual into new tissue or energy. The metabolic notion earns its theoretical keep by playing a key role in a broad story about the growth and self-maintenance of biological form.

An alternative concept of life, whose task is to shed light on evolution, requires the possession of a genetic code. A living thing possesses machinery that enables it to reproduce itself. Only through gradual modification of this potentially immortal method of storing information can complex life forms have come about. Here, *living being* will be synonymous with *gene bearer.*

Although almost all the individuals that common sense calls alive are included in the overlap between these two naturalizations, there are certainly areas of disagreement. A virus has genes but no independent metabolism. The genetic conception counts it as alive, but the metabolic conception does not. What would we say about the last dinosaurs if we learned that their demise was occasioned by an intense burst of solar radiation that destroyed their germ-line DNA? This last generation will miss out on being alive according to some versions of the genetic view. Despite this, they are paradigms of life in the metabolic account.

The emerging area of Artificial Life makes a further demand on the commonsense concept.[2] Many claim that the self-replicating, self-organizing structures generated in computers are more than just simulations of life. This claim requires a notion that abstracts away familiar ideas about how life is physically constituted.

The task of laying all proposed accounts side by side and deciding among them might be worth the effort if we believed there to be some deep life-fact that is a target of our definitional efforts. However, we saw that in a physicalistic universe there is no place for most popular conceptions of the deep life-fact. The cleanly naturalistic conditions sketched above are much more interesting to science than any multiple-qualified set of statements that might jointly constitute the commonsense concept. Common sense does not drop out of the picture altogether. It keeps the new notions on a leash, however lax. They must share enough with its commonsense ancestor to merit the commonsense title *life.* When the class of objects that a proposed naturalization picks out does not usefully overlap with the class isolated by common sense, we should recommend that the specialist find a new term.

If an area within the biological sciences can have a concept of life fine-tuned to suit its needs, then why shouldn't ethical theory? The notion I am looking for will bring to the surface a range of morally interesting properties scattered throughout the natural world. In building such an account, I must meet three sets of demands. First, like the above scientific stories, the notion must keep some faith with the commonsense conception in order to make a claim on the term *life*. The second need is that the notion be morally interesting. The metabolic and genetic accounts build on intuitions about life in a scientifically helpful way by making explicit links with specific areas of biological theory. To help the moral philosopher, we need to adapt common sense about life so as to establish clear links to familiar normative notions. Third, the account must be consistent with what science tells us about the natural world. Accounts of the value of nature prominent in some traditional societies have fallen foul of this requirement. The attribution of souls, life force, or inner spirits to moas and kauris may be consistent with a folk understanding of life; it would make up part of a story about how intuitively living things manage to self-move. Further, the intentional states of these spirits would make them worthy bearers of value.[3] Modern science, however, rules out their existence.

A BIOFUNCTIONAL EXPLANATION OF SELF-MOVEMENT

We need a theory that bridges the gap between life and the psychological view about intrinsic value. The biofunctional-movement conception of life that follows attempts to explain in naturalistic terms the self-movement that we identified as central to the ordinary-language notion. This leads to a representational conception constructed, in part, out of customized descendants of folk psychological concepts.

A simple evolutionary story will introduce these accounts of life. The earliest biological or prebiological entities would have been unadorned self-replication machines. They existed in seas filled with organic compounds consumed to allow self-replication.[4] There would have been no structure whose specific job was to move them around their environments, and random motion generated by aquatic currents would have sufficed to keep them in touch with the raw materials for self-replication. These replication machines are paradigms of life in the genetic account. However, regardless of any genes, or any genelike devices, the folk account has difficulty acknowledging them as alive. Their movements are, in almost all circumstances, identical to those of clearly nonliving matter. They do not move themselves.

As the seas filled with self-replicating structures depending on a finite supply of organic compounds, selective advantage would have accrued to those organisms able to compete more efficiently for the raw materials of replication. One evolutionary strategy would have focused on movement devices. The parts of any object can move it in specific ways when acted on by external forces. How do we distinguish those parts of self-replicating structures whose function is to promote self-movement? The biofunctional-movement account selects from among those things that the genetic definition counts as alive. It defines life as follows: *Something is alive according to the biofunctional-movement account if it has a state whose biofunction is to produce specific changes to, or movements of, it.*

The biofunctional-movement account recognizes as alive things with devices whose purpose is to help them resist or modify external forces. A structure whose biofunction is directed at activities other than self-movement or self-change may have the effect of moving or modifying their bearers. However, this will only count as a by-product of the genuine selected function. We can assign a structure the biofunction of movement, in accordance with the etiological theory of function, if it exists because earlier copies moved its bearer around its environment in a fairly specific way. Perhaps some of the earliest movement devices might have been like the flagella of certain bacteria, propelling their bearers with whiplike actions. These organisms' paths through the sea would have been quite different from those of non–self-replicating objects of roughly the same physical dimensions.⁵

I have offered a naturalistic explanation of the self-movement or self-change we intuitively associate with living things. In adding "self-change" to the definition, we can account for plants, most of which change their shape and size in order to bring materials for survival and reproduction within reach, rather than moving their entire bodies.

I return to my evolutionary story. The capacity for movement marks one important boundary. Later, another innovation would have arisen. Organisms would have acquired the ability to produce movements, not just randomly or all of the time, but in response to fairly specific environmental prompts. We can call the structure linking the sensor of the ecologically salient environmental property, to the movement producer, a *representation*. A representation is a device that contains information about the environment. Biofunctional explanation will type representations in terms of what they are supposed to do. A representation is a structure whose biofunction is to appropriately modify or funnel the impact of environmental forces through to movement or change. Such movement or change will be in response to the environmental property whose name figures in the content of the representation. No great biological sophistication is required. Bacteria have sensors attuned to such things as heat and light and certain chemical markers, and we can find representations in even simpler things. The virus T4 replicates with the help of a bacterial host.[6] It possesses receptors that bind to sugary substances on the bacterium's surface. When this happens, a response is triggered and the genetic material of the virus is injected into the bacterium, where it hijacks its host's replication machinery. This very basic description makes it clear that there is a structure in the virus selected for both its sensitivity to a certain environmental condition and its triggering of a specific movement.

Many of the things that the biofunctional-movement account counts as alive possess structures falling under this liberal concept of representation. It is central to the ethical theory that I will advance, and so as to isolate the large subset of things alive according to the biofunctional-movement account that are also representers, I offer a new definition of life. This is the representational conception.

Something is alive according to the representational account if it has a state whose biofunction is to produce specific changes to, or movements of, it in response to particular states of the environment.

The relationship between representation and biofunction has been the focus of much philosophical interest. The teleological theory of content due to Ruth Millikan, Colin McGinn, and David Papineau, among others, sets out to show how mental meanings could be part of the natural order by reducing the semantic properties of human beliefs and desires to their biofunctional properties.[7] A desire for cake has the biofunction to get its bearer to consume cake. A belief that there is cake can perform its biofunction only in a cake-containing environment.[8]

This theory is extremely controversial. One line of criticism has fixed on what kind of content emerges. Jerry Fodor thinks that it is irredeemably disjunctive due to what he perceives as an ineliminable indeterminacy in the description of natural selective forces.[9] Natural selection cannot help us in choosing between an array of alternative content ascriptions. Take the above story about the representation in T4. According to Fodor, the teleological theorist must agree that it is as right to say that T4 represents "bacterium," "sugary bacterium surface," "ideal host for DNA," or so on. As intentional content does not exhibit these radical indeterminacies, Fodor concludes that the teleological project fails.

A number of writers have responded to Fodor on his own terms, arguing that this indeterminacy can be cleared up.[10] In chapter 6 I will show that the biofunctional account can give us more content specificity than

Fodor supposes. It will not bother me if this specificity is insufficient to defeat Fodor's attack on the biofunctional reduction of folk psychology. My interest in the biofunctions of representations is motivated by concerns different from those of Millikan, McGinn, and Papineau. Rather than trying to show that everyday content is teleological content, I am interested in explaining the directed self-movement that emerges in relatively simple living things.

Fodor's argument, if successful, shows that it will be difficult to reduce determinate folk psychological content to indeterminate disjunctive biofunctional content. It will not show that states that guide a nonhuman individual's movement cannot be understood in biofunctionally contentful terms. The fact that there are many ways to describe causally relevant objects in an organism's historical environment does not cancel the utility of an explanation pointing to some or all of these descriptions. We should require that each disjunct point to a causally relevant factor in the evolution of the representation. Some characterizations will point to so few causally relevant factors as to be next to useless. More extensive life-representational ascriptions will enable us to tell a more complete story about the causal/selectional history of the representation.

Rather than being a scientifically cleaned-up replacement for folk psychology, the teleological theory is supposed to complement our folk story about meaning. Folk psychology did not evolve to meet any need to explain or predict stick insects, so it is not that surprising that the theory struggles with nonhumans. The teleological theory kicks in where folk intuition begins to lose its grip. It extends content explanation beyond a narrow historically favored group and thereby paves the way for the moral appreciation of nature.

Explanatory strategies inappropriate to the nonliving become available for the representationally alive. Content is central. A stone has no content-characterizable goals. It has no structure whose biofunction is to produce movement or change. The lack of content-containing states has consequences for the explanatory strategies that will tend to work for the rock. It not only obeys the laws of physics but also tends to act in accordance with the very simple principles of folk physics out of which we build our intuitive theory of the nonliving world. External forces can be fairly easily traced through the rock to movement. Though, of course, representing organisms are as constrained by the same real laws of physics as any material thing, their possession of movement devices makes the task of explaining the movement of an object in terms of detectable external forces more difficult. Even when we are in a position to open an organism up, deciding how internal mechanisms maneuver their bearers around the world promises to be a very difficult task. Fortunately, the impact of the internal state of T4 on its movement is explainable in terms of relations to selecting properties in its environment. Biofunctional representers possess goals directed toward the environment. The moral account I go on to outline makes much of these goals. To prepare the way for this moral case, I refer to them as *biopreferences*. The goal of a representation will be the relationship between its bearer and the key bit of the environment that it was selected to bring about. Given Fodor's indeterminacy point, we would characterize the biopreference of T4 as perhaps "injection of material into bacterium" or "lodgement of DNA in sugary-coated thing" or "replication of genes," and so forth. Any content ascription that brings with it new causal/selectional information improves our ability to predict how inner states will affect the movement of an organism. I will offer more detail on these contents in chapter 6.

WHY THE REPRESENTATIONALLY ALIVE ARE MORALLY IMPORTANT

Notions of representation, content, and goals can be useful when we come to simple living things. We know that some explanatory tasks are transferred from familiar folk psychological content to the new biofunctional notion. Can associated value also be transferred? To many who have followed my discussion up to this point, the notions described above will seem nothing more than normatively inert simulacra of our familiar value-endowing concepts.

If the general account defended in chapter 3 is true, the fact that biofunctional content is not identical with folk psychological content will not be an obstacle to its moral importance. A state can be a plausible naturalization of a folk psychological state without being identical to that folk psychological state. If life-constitutive content is as explanatorily useful as I have suggested, we could consider it a plausible naturalization of familiar folk psychological content. All the representationally alive are, therefore, candidates for some value attaching to the ordinary-language notion. I propose an identification of an organism's moral good with its content-characterized biopreferences. In pushing a boulder down a hill, we interfere with no goal, and therefore no good, of the boulder. Stepping on a cockroach approaching a food scrap is a quite different matter. The cockroach possesses environment-directed internal states whose biofunction is to enable it to retrieve the food scrap. The notion of a content-containing goal enables us to see how the environment can be rearranged so that it is wrong for the cockroach.

This account supports a moral distinction between the living and the nonliving. The task is now to make appropriate discriminations among the living. I suggest that highlighting the relationship between each living thing and value-endowing folk psychological notions allows us to assign value broadly but differentially throughout nature.

Living things can be placed on a continuum starting with the most representationally simple and moving up to the most complex. T4 seems only to have one representation, which is sensitive to the sugary coating on bacterial surfaces. It seeks bacteria, or sugary surfaces, but nothing else. Further, its internal structure permits only one channel from the triggering of the representation through to movement. Hence the virus is capable of doing only one thing: injecting its genetic material into the prospective host. An organism that has representations sensitive to a wide range of environmental properties can deviate from the mandates of its environs more often than its comparatively impoverished cousins. There will be more environmental keys triggering movement, and there may well be a distinctive movement for each key. Representational complexity can be relevant to goals in other ways. Simpler organisms form goals pretty much spontaneously in response to specific properties in their immediate environments. Others integrate information from many sources. They can form representations about, and thereby have goals concerning, things that no longer exist or are thousands of kilometers distant.

Why should this increasing representational complexity be of interest to the moral theorist? Remember that where we are disposed to assign folk psychological states we are also disposed to assign value. If we acknowledge folk psychology's value-anchoring role, we can see the moral importance of greater representational complexity. Placing organisms on the above continuum gives a reasonable measure of how much of folk psychology can be used to predict and explain them.

The goals of simpler organisms will be fairly straightforwardly biofunctional. Biofunctional states bring certain practical difficulties. One of the more awkward features of biofunctional content ascription is its historical nature. To make an accurate ascription of biofunctional content, we must know about the selectional history of the representation in question. If we do not have the detailed historical information to make accurate biofunctional ascriptions, we can appeal to fragments of folk psychology to do the required explaining work. When we assign to the virus a desire for bacteria of a certain type, or the belief that there is a bacterium of this type in the vicinity, we use folk psychology as a guide to biofunctional content. Of course, if we accept Fodor's point, it may be equally appropriate to assign beliefs and desires about sugary surfaces. While fully biofunctionally contentful, T4 turns out to be only marginally folk psychologically contentful.

Failure to accord moral consideration to organisms possessing only representations that are the targeted results of natural selection leads to unwanted consequences when we turn to human beings. Some writers have attempted to explain a wide range of human cognitive abilities selectionally.[11] The success of the program of evolutionary psychology should not make us retract claims about the moral worth of some human acts. Even if the program fails as a whole, there are areas of human mentality for which it should prove fruitful. Though in us there is a substantial cultural overlay, parental interest in the welfare of offspring will have a significant component that is essentially the same as in chimps, cats, and even bees. Effective care for young must have been central to our evolution. We certainly do not think that interest in the welfare of one's children ought to rate lower than concern for a phonecard collection because the former has a more obvious evolutionary explanation than the latter. Representations can confer goods on an organism if they are generated in their bearer's lifetime, or if they are the results of intergenerationally operative natural selection.

Greater internal complexity often makes folk psychology more useful. But not always. Some things on this planet are more cognitively complex in ways that cannot be explained by folk psychology. Much of the information-processing effort of the brain of the African longnosed elephant fish goes into detecting and analyzing disturbances in its electrical field. The fish certainly has internal states that are selected to deal with external conditions. However, its distance from us means that our commonsense psychological theory cannot always provide accurate surrogates for these states. So, folk psychology's value-anchoring role will be bad news for the longnosed elephant fish. It will not be as morally valuable as other organisms of similar representational complexity.

According to the view I am describing, our value assignments track the applicability of value-anchoring folk psychological talk. The representationally alive become candidates for value because they possess states properly described by conceptual descendants of our ancient value-anchoring theory. How much value should be assigned? We observe that as goal-relevant cognitive complexity increases, movement-directing states tend to approximate more closely folk psychological paradigms. And so we spread value broadly but unevenly throughout the living world.

Does it really make sense to assign even a miniscule amount of intrinsic value to things of T4's simplicity? Natural properties do not always come in neat packages attuned to human perceptual and behavioral abilities. For example, some foods contain amounts of dietary fiber that the sensible nutritionist should treat as nonexistent. Yet the fact that it is rational for finite digesters to treat these foods in this way does not stop them from being as determinately fiber-containing as apples or oats. That life-representational value can come in amounts too small for humans merely emphasizes its naturalness. Though a single representationally simple organism may not qualify for a place in the human moral calculus, we may allow that large numbers of that organism might. When humans destroy an ecosystem, a very large number of organisms of varying representational sophistication are affected. Correspondingly large numbers of goals and goods are interfered with. We must balance this effect against the frustration of the often relatively few and unimportant goals of relatively few humans.

We should note that some human goals conflicting with those of nonhumans are quite peripheral.[12] By this I mean that not fulfilling the goal has little impact on the ability to satisfy other goals. A reduced bluefin tuna quota may mean that the dietary preferences of some humans are not satisfied. In many cases, there will be alternatives that contribute almost equally well to the goal of being adequately nourished and able to pursue other goals. Matters will often be quite different when we look at nonhumans affected by our actions. The goals we interfere with here will be quite central rather than peripheral. Obstruction of the goal in question will lead directly to the frustration of a range of other goals. Overfishing of bluefin tuna thwarts almost all the goals of those individuals that are killed.[13] Further, as the population plummets, surviving tuna will be much less likely to satisfy other mate-finding goals.[14]

OTHER- AND SELF-DIRECTED GOALS

I turn now to a range of apparent objections to the life-representational ethic. Standard talk of representation takes individuals as given. The things that we are most interested in explaining are single humans and some nonhuman individuals. In using representations as markers of moral worth, I cannot afford this complacency, however. If value flows from representation, then all representers will be morally considerable, not merely some subset that most engages our interest. It may be strange to think of viruses as representing their surroundings. Yet there are cases included by the representational account that are even more scandalous to common sense and to any moral theory than T4.

The commonsense notion of mental representation recognizes one representer in my body—me. According to the liberal biofunctional account, beyond myself are all the bacteria and many of the viruses in my gut and bloodstream. Even more unexpectedly, any part of my body that depends on some signal to carry out its selected biofunction will be a representer. Each will presumably possess its own goals and goods. One of the liver's many functions is to remove excess glucose from the blood and store it as glycogen. When blood-sugar levels fall, glycogen is reconverted into glucose and pumped back into the bloodstream. Take this second function of the liver: the transformation of glycogen back into glucose. This will count as representation-guided behavior according to the biofunctional account; there is a structure within the liver that is selected for its sensitivity to low blood sugar and its propensity to cause glucose-manufacturing behavior. Does this mean we must assign independent value to representing livers?

I now show that the correct understanding of biofunctional contents enables us to exclude parts of organisms as candidates for separate intrinsic value.

The first move is to distinguish two types of goals. A goal is *other-directed* if its contents refers to the goods of objects outside the bearer of the representation. Only the good of the representer will feature in *self-directed* goals.

The liver's goal is other-directed. This other-directedness becomes apparent once we examine the relevant selection history. The representation in question depends on past copies of those representations having guided past livers in boosting the reproductive chances of the possessing organism. Had past copies of the representation not enabled past livers to contribute to fitness, there would be no such representations today. This means that the content-characterized goal of the liver will make reference to the survival of the organism that contains it.

Now, a narrow characterization of the goal may omit all references to the good of the possessing organism, perhaps talking only about the transformation of glucose into glycogen and back again. Such a story is rather incomplete, failing to give much causal/historical information. Further, these narrow goals are teleologically subservient to wider, other-directed goals. The transformation of glucose into glycogen exists to promote human fitness.[15]

The possession of an other-directed goal has important consequences for the correct treatment of the liver. Under normal circumstances, destruction of the liver undermines its good. Liverless humans do not generally fare well. However, if the liver is replaced by an equally efficient organ, this good may not be harmed.

Similarly, other-directed characterizations of the goals of individual organisms are often much less appropriate. Though a koala may depend on a given gum tree for its survival and self-reproduction, its representations have not been selected to maintain that particular, or any other species of, tree. The koala's good certainly depends on the gum tree, but is not to be characterized in terms of it. As it captures no causally relevant information, "gum tree preservation" will be a spurious addition to the content of the koala's goal. Though some of the koala's goals will be other-directed—targeted for example, at the goods of offspring—we are unlikely to find any directed at the good of the gum tree.

As chapter 7 will show, certain individual organisms do have substantially other-directed goals. Worker ants are as teleologically subservient to their colonies or queens as my liver is to me. This means that moral interest in ant representations will draw interest away from the ant and toward its colony. Some of the biopreferences of organisms that have long coexisted with specific environments may turn out to be partially other-directed. They will have been selected, at least in part, for their role in preserving a given environment. This environment will, therefore, make up part of their good, and focus on representations establishes the moral importance of the collectives or wholes that preoccupy many environmentalists.[16]

I pass now to a second category of problem objects. The representational account of life I offered above demands some manner of evolutionary history. This means that it does not count human-designed objects as alive. Now, the job of the evolutionary story about content has been to show that kinds found in simple living things are relatively close descendants of certain of our familiar value-anchoring notions. It would be a mistake to use the account to exclude nonbiological representers. The human-endowed functions of a silicon brain may enable its bearer to form plans and, perhaps, entertain conscious thoughts.

It would seem that we do not have to await the construction of machines as complex as *2001*'s HAL 9000 and *Star Trek*'s Data to find machines that might be candidates for value, however. I have argued that biological things much more representationally basic than humans can be morally considerable. Should we say the same for many of the unsophisticated mechanical devices that already populate the world? We would need to include such primitive representational devices as thermostats, pocket calculators, and gasoline gauges.

The preceding discussion indicates how to avoid assigning independent moral value to these things. The representations of thermostats, hand calculators, and gasoline gauges will tend to have functions that, like those of hearts and lungs, are teleologically subservient to human goods. A thermostat may represent the temperature in a room, but an accurate description of functional dependencies allows us to recognize that it does so to serve human ends. "Maintenance of human-friendly temperature," or similar, will be indispensable in our content characterization of the other-directed goal of the thermostat. It also captures the good of the thermostat. Widespread destruction of thermostats will generate moral wrong only if other devices capable of maintaining temperature for us are not installed.

As mechanical representers become more complex, they may begin to acquire a degree of teleological independence. Such devices will generate goals with self-directed contents. These representations will, perhaps, guide the movements of the machine so as to preserve its ongoing existence and acquire the wherewithal to pursue its own goals. At this point we may indeed find representing machines with values similar to those of simple living things.

This chapter has shown that our entrenched, apparently anthropocentric moral views can take us some distance into nature. The representational account of life acts as a bridge between living things and value-anchoring psychological notions. It enables value to be spread very broadly throughout nature. Individual things are not all to be valued equally, however. The amount of value we assign to an individual depends on the range and complexity of the goals of which an organism is capable. Why does this type of complexity matter? As organisms have more varied and numerous goals, they tend to become more folk psychological. Folk psychological notions, in turn, have the closest association with relevant normative notions. Thus the life-representational ethic both acknowledges the preeminent place of humans on this planet and spreads value broadly enough to provide at least the raw materials for an environmental ethic.

CHAPTER 6

꿈

The Contents of Biopreferences

THE TELEOLOGICAL ACCOUNT OF CONTENT

The focus on biofunctionally characterized representations gives us an account of life suitable to serve as a bridge between entrenched moral commonsense, as embodied in the psychological view, and parts of the inanimate world. Living things are valuable by virtue of their representationally characterized biopreferences. The notion of a biopreference is obviously an important one, and it will be the primary focus of this chapter.

I require a concept sufficiently and precisely specified to perform two tasks. First, the sophistication of actual and potential biopreferences will give a measure of the value of organisms. Second, the contents of biopreferences are to guide us in our treatment of organisms.

This chapter responds to two morally dangerous implications of the teleological approach to biopreference content, tracing a path through recent debates in the philosophy of mind and in the philosophy of biology.

I have touched on one dangerous implication in the last chapter. Jerry Fodor claims that facts about selection history fail to make representational contents determinate. Now, a little indeterminacy may not be such a bad thing when our focus is a very simple organism; overly specific content for a bacterium is surely only an artifact of analysis. However, the moral theorist cannot countenance too much indeterminacy. The correct treatment of an organism needs to be sensitive to its goals. Unless we can make these goals moderately determinate, it will be impossible to say what treatment is consistent with the organism's goals and what treatment frustrates them. Indeed, if the biopreferences of stick insects turn out to be radically indeterminate, then the obvious conclusion will be that *any* treatment will conform with them. Though the argument in chapter 5 may have shown stick insects to be morally valuable, it will turn out that they are valuable but only in such a way that is irrelevant to human choices. I aim to meet this worry by showing that biofunctional contents can be made quite determinate.

This leads to further difficulty. In chapter 4 I have pointed to the problem of genic selectionism for those who would inject biological teleology into an account of individual value. The genic selectionist holds that the goal of any biological structure is most fundamentally the replication of underlying DNA. If this thesis is true, an ethic built on selected goals will privilege not individual organisms, species, or ecosystems, but genes: a highly unwelcome outcome for an environmental ethic. This chapter will show that explanations of selected purpose are better conducted in terms of phenotypic characters than in terms of genes. The important implication is that an ethic grounded in selected purpose need not privilege genes.

With the notion of a biopreference more precisely delineated, I go on to compare an ethic founded on it with more familiar views that trace a nonhuman organism's value to its sentience.

The Problem of Content Indeterminacy

Two types of contentful states figure in the representational conception of life. Using folk psychological taxonomy as a guide, we can divide these into belieflike states and desirelike states, each with characteristic biofunctions. Belieflike states have the function of carrying information about key chunks of the organism's environment. When near to a certain type of bacterium, T4 needs to have a state containing the information "bacterium" or perhaps "sugary surface." The desirelike goals or biopreferences have been my chief focus. Their function is to bring about a specific relation between an organism and its environment. The last chapter has shown that candidate content ascriptions to this type of state in T4 are "injection of material into bacterium" or "lodgement of DNA in sugary-coated thing" or "replication of genes." Though these states are certainly plausible naturalizations of "belief" and "desire," they differ from their folk psychological ancestors in important ways. For starters, in humans, beliefs and desires are separate states. The two varieties of function described above may be subserved by a single state only in cognitively simple organisms.[1]

The last chapter has drawn extensively on the teleological theory about human intentionality due to Millikan, McGinn, and Papineau. This theory was inspired by the need to show how mental meanings could be part of the natural order. Its advocates hold that function or purpose determines the semantic properties of beliefs and desires. This story counts as naturalistic because these functions or purposes are accounted for as the products of natural selective forces rather than of, say, divine design.

Critics complain that contents assigned by the teleological theorist often look rather different from those we intuitively assign. Two kinds of objection have been prominent in the literature. The most striking revolves around Davidson's Swampman, discussed in chapter 3. Though the representational states of Swampman have no selectional histories, we can perfectly well use folk psychology to account for its behavior. This seems to make Swampman folk psychological without being biofunctional. The second objection is that of content indeterminacy. We saw that according to Fodor, the radical indeterminacy of teleological contents makes them fundamentally different from the rather determinate folk psychological contents.

How worrisome are these objections? In chapter 5 I have pointed out that my interest in the teleological theory is motivated by an entirely different set of concerns from those of Millikan, McGinn, and Papineau. I do not want to show that everyday content is teleological content. This change in orientation has important implications for our treatment of the objections to the teleological theory. If our aim is to show that folk psychological meaning turns out to be teleological meaning, we will be concerned about the assignment of contents that differ markedly from those sanctified by semantic intuition. If we want, instead, to assign contents to things beyond the traditional scope of folk psychology, we will be asking whether the theory tells us how to assign contents appropriate for moral theorizing.

The new focus leads us to look differently upon the two objections. Swampman crystallizes strong intuitions about the supervenience of folk psychology on current behavior. Though it may show that the inner states of swampmen, swampmonkeys, and swampmice are less plausible naturalizations of folk psychological notions than states of their etiologically conventional counterparts, it does not challenge the teleological theory in its job of pragmatically extending content explanation to nonhumans. The objection, therefore, does not threaten to rid our teleological account of moral interest. In contrast, the indeterminacy argument undermines the teleological account in this moral-theoretic role. The faults that Fodor identifies cannot be traced to any unanalyzed prejudices about meanings; rather, they threaten to make the theory unusable by the moral philosopher.

Fodor's Objection

Our investigation of indeterminacy begins with a much-discussed story. The frog's visual system is designed so that a fast-moving dark object flying through the visual field will trigger a certain response; the frog shoots out its tongue and attempts to capture the object. In the natural environment of the frog, these fast-moving dark objects are almost always flies, and this is convenient for the frog because as a result of its efforts, it gets a meal.[2]

It seems likely that there is some inner state of the frog, call it *R*, that mediates fly perception and tongue snapping. How do we assign content to *R?* Many writers have been attracted to some version of the indicator theory of content, originally championed by Fred Dretske (1981, 1988). They hold that content is to be assigned on the basis of information carried. The inner state carries information about flies because its tokening is reliably correlated with flies crossing the frog's visual field; therefore, the state has content "fly."

Matters become more complicated as the story continues. Frogs are famous for the ease with which they are fooled. An air gun pellet hurled at more or less the appropriate speed across the frog's visual field will provoke the same response. As *R* is reliably caused by air gun pellets, it seems that *R* will carry information about air gun pellets as well. Does this mean that the content we must assign to *R* is not "fly" but "fly or air gun pellet"? If so, as there are always going to be situations in which the most bizarre things reliably cause *R*, parallel reasoning will lead us to the conclusion that the frog can never err or misrepresent.[3]

The teleological theorist hopes to offer a way of distinguishing between genuine representation and misrepresentation. The frog possesses a structure that "represents *when the token is caused by circumstances of the same kind as those selectively responsible for the existence of the type.*"[4] Thus, the frog errs when *R* is caused by circumstances similar to those for which *R* is selected. At first glance, the frog will represent accurately when a fly causes *R*, but not when an air gun pellet does.

Fodor (1990) has challenged this conclusion. He thinks that an ineliminable indeterminacy in the description of natural selective forces makes teleological content disastrously disjunctive. There are many equally adequate ways of describing the situations that historically lead to the existence or proliferation of the type, and so natural selection cannot help us in making the choice between an array of alternative content ascriptions. Fodor says that "just as there is a selectional explanation of why frogs should have fly detectors, . . . so too there is a selectional explanation of why frogs have little-ambient-black-thing-detectors. . . . The explanation is that *in the environment in which the mechanism Normally operates* all (or most, or anyhow enough) of the little ambient black dots are flies."[5]

If Fodor is right, the teleological theorist is frustrated in the desire to assign univocal content to R. Teleology says nothing to the issue of whether R should be mapped onto flies, some environmental property contingently associated with flies, or even states of the frog's perceptual apparatus and, therefore, cannot exclude air gun pellets, distant 747s, or retinal blobs from the content of R.

Though we might tolerate some indeterminacy in the biopreferences of simple organisms, we cannot let this indeterminacy get out of control. Goals are supposed to guide us in our treatment of an organism. It is by appealing to these goals that we understand which are the right ways and which are the wrong ways for an organism to be related to its environment. If the teleological theory tells us that any characterization is as good as any other then, we learn either that there is no way to mistreat the organism or that we cannot use teleological contents to understand what counts as correct treatment of the organism.

I suggest that the examination of selection forces does enable us to eliminate much indeterminacy. Though the residue may be sufficient for Fodor's condemnation of the teleological theory as a reduction of folk psychology to stand, we can rescue the distinction between the morally correct and morally incorrect treatment of a nonsentient organism.[6]

Let's return to the fly-eating frog. We know that some kind of selectional story accounts for the origin of R; R has been selected to equip the frog to deal with some property related, perhaps far from universally, with flies. Attention to relevant counterfactuals permits a fine-grained description of R's selection history and the forces that have sculpted it. Fairly specific content will emerge from this description. The idea is quite simple. If the modification of a certain environmental property in relevantly similar, biologically consistent, counterfactual histories changes a particular aspect of the structure of R, we can say that it is selectionally relevant to R.

We can make a few guesses about which properties will have been selectionally relevant to specific components of R. For example, there is likely to be a piece of frog representational machinery belonging to R that depends on the statistically normal angle of flight and velocity of historical flies. This component of R will have the job of causing an appropriately timed and directed snap. Facts about the darkness of historical flies, relative to normal backgrounds, will have had important implications for aspects of *R* most proximally connected to the frog's visual system. The nutritional properties of the fly will be selectionally responsible for, among other things, the part of *R* that produces mouthward movement of the tongue after a successful strike. In juggling the properties in counterfactual histories, and noting the persistence or lack of persistence of structure, we establish which representational features depend on which environmental properties. Certain properties will likely be excluded. If we try this test for such properties as the fly's precise genetic makeup, or its degree of relatedness to organisms belonging to other species, the answer will be different. Changing either of these need not result in any representational change.

Some will think that I am making questionable assumptions about the evolution of *R*. Stephen Jay Gould and Richard Lewontin (1984) famously criticized the tendency of adaptationists to look upon every aspect of any organism as exquisite design. They counter that natural selection does not have entirely free run of the biological world. The availability of materials, random genetic drift, and design compromises all constrain selection. Our account does not need to be so extremely adaptationist as to assume that all the structural elements of *R* can be traced to properties in the environment that are selectionally responsible for them, however. We can allow that there may be behavior-causing components that have not been explicit targets of selective forces.[7] These features may be spandrels, or side effects, of selection for some other property. The teleological approach I am sketching is powerless to account for behavior caused by representational elements produced in this way, so I hope that the majority of behaviorally relevant representational structures does not fall into this category.

Fodor thinks that teleological content is bound to be messy and disjunctive. On the contrary, the content that emerges from due attention to selective forces will be messy-looking and *conjunctive*. The names of all the properties that have been selectionally relevant to *R* need to be included in our content ascription. *R* will be assigned content that is something like "dark and small and fast-moving and nutritious and. . . ." Not all of the names included in our conjunctive account of *R* will pick out properties of the fly itself. For example, facts about the darkness of the background against which historical flies have flown will have been evolutionarily important, and this importance will be mirrored in the structure of *R*.

The above teleological content is untidy. However, we should recognize that any queasiness about this stems from folk psychological prejudices about the way contents ought to be. What at first seems a disordered conjunction of folk psychological contents is actually the most efficient way of picking out a determinate set of selectional/counterfactual dependencies between the frog's inner state and its environment. A full statement of the teleological content of a representation will be a detailed story about the properties that have historically been of high ecological salience for organisms belonging to that species. In chapter 5 I have claimed that we could find folk psychological surrogates for many selected goals. Any decent folk psychological surrogate for the state will contain many of the conjuncts that pick out properties grounding these selection forces. Frogs will need, somewhat clumsily, to be described as desiring each of the properties picked out by the conjuncts.

Messy conjunctive teleological content begins to look sadly out of place when we turn to the representing states of organisms such as humans. A great deal of work can be saved by pointing to differences between my explanatory purpose and the purpose of Millikan and Papineau. For Millikan and Papineau, the teleological theory is called upon to reduce human intentional states. They bother with simple representers only as a means to this grander reductionist aim. Were either of these writers to accept my rendering of teleological content, they would need either to show how complex representers might have less conjunctive contents or to argue that human contents are secretly very conjunctive. In contrast, the teleological theory is *not* for me a reducing theory. I am prepared to accept folk psychological contents at face value, appealing to them to understand a person's good. Rather than displacing folk psychology, the teleological theory steps in where folk psychology's grip on representation begins to weaken.

THE THREAT OF GENIC SELECTIONISM

On to a further problem. We may have alleviated the problem of indeterminacy only to find the teleological theory assigning contents entirely inappropriate for an environmental ethic. Difficulties arise when we ask what kinds of biopreferences organisms will tend to have.

In chapter 3 I have briefly considered the prospects for an argument that would make genes intrinsically valuable. Things that contain plausible naturalizations of value-laden psychological notions are candidates for moral worth. Selfishness is a commonsense psychological notion, and Dawkins suggests genic selfishness as a plausible naturalization of it. I have explained why Dawkins's thesis holds no terrors for moral theory. Genic selfishness turns out to be a rather implausible naturalization—it is quite distant from its commonsense ancestor, meaning that a correspondingly small fraction of any original moral load will be transmitted. Moreover, a quick examination of the psychological view reveals that selfishness in isolation does not carry much value. So this argument for moral gene-centrism is turned back.

An interest in teleological contents may drive us once again to genic value. Seemingly paradoxically, by placing great importance on the representational states of individuals, we may find that the genes of these individuals, rather than the individuals themselves, are what matter morally. In the preceding section I have explained that, by tracing them back to properties that have done selecting work, we can make the contents of selected representations relatively determinate. An exact story about the content of an individual organism's goal emerges from an account of the aspects of the historical environment that have done causal work in originating and maintaining the state underlying the goal. Williams (1966) and Dawkins (1989, 1990) have argued that any selected purpose in nature is most fundamentally directed at the preservation of DNA. This will have unwelcome consequences for a theory that equates respect for an organism with respect for its selected goals. An interest in biopreferences will combine with an explanatory gene-centrism to yield a moral gene-centrism.

Perhaps this DNA-centrism is not so scary after all. DNA-directedness might be taken as the feature of innate and simple, naturally selected goals that distinguishes them from goals formed in their own lifetimes by more complex, cognitively plastic organisms. Environmentalists might usefully point to DNA-directed goals to help justify their attempts to maintain breeding populations. Though we may be able to guarantee each individual belonging to the last populations of tuataras a long and pain-free life in separate zoos, it is only in keeping species-mates together and providing conditions suitable for breeding that we can ensure that any DNA-directed goals will be satisfied.

Despite these pragmatic benefits, it is an unattractive feature of an ethic that it assigns intrinsic value to sequences of nucleotides rather than to individual organisms. But DNA-centrism is not an enforced outcome of close attention to selected goals. To understand this, I will take a close look at the arguments for explanatory gene-centrism, or genic selectionism.

Dawkins and Genic Selectionism

Genic selectionism is a claim about purpose in nature. Dawkins, the best-known modern advocate of the view, contrasts the explicitly teleological language that we must use to discuss phenotypic characters—"Birds' wings are obviously 'for' flying, spiders' webs are for catching insects, chlorophyll molecules are for. . . ."—with the approach we must take to DNA: "What *are* DNA molecules for? The question takes us aback. In my case it touches off an almost audible alarm siren in the mind. If we accept the view of life which I wish to espouse, it is the forbidden question. DNA is not 'for' anything. If we wish to speak teleologically, all adaptations are for the preservation of DNA; DNA itself just *is*."[8]

Dawkins's argument that the gene is the unit of selection hinges on the claim that genes are Active Germ-line Replicators. For Dawkins, a replicator is "anything in the universe of which copies are made."[9] There are two axes along which replicators can vary; they can be *active* or *passive*, and *dead-end* or *germ-line*. An active replicator is one "whose nature has some influence over its probability of being copied."[10] This Dawkins contrasts with the passive replicator, like a sheet of Xerox paper, which exercises no such direct influence.[11] In order for a replicator to be germ-line it must be "potentially the ancestor of an indefinitely long line of descendent replicators."[12]

Thus far, there is an analysis of the concept of the unit of selection. We must now ask which entities in the natural world meet Dawkins's requirements. Individual organisms do not qualify. Though they seem to be active—what individuals do is evolutionarily important—organisms are not replicators. It is most important that changes to replicators be passed on to the next generation. The blending of genetic heritages ensures that the offspring of sexually reproducing organisms do not exactly resemble either parent. Asexual organisms may not mix their genes in the act of reproduction, but they still fail to be replicators. The injured aphid differs from the mutated gene in not passing on its new properties to succeeding generations of aphids. So no organism, regardless of how it reproduces, is an active germ-line replicator.

In contrast to organisms, genes perform well in the role of replicator; if they evade various genetic-repair mechanisms, mutations to DNA do reliably turn up in later copies of DNA. In what follows I will turn the spotlight to the other side of Dawkins's analysis of "unit of selection"—the activity requirement. The question is: do genes affect their surroundings in such a way as to promote their replication prospects? Since genes do not directly confront their environments, they can be active only by working through the bodies of organisms.

Dawkins needs to tell a story about how genes manage to exercise systematic influence by means of the bodies of individuals. To this end he proposes a causal analysis whose task is to provide principled grounds for mapping genes onto parts of the housing organism, or conventional phenotype, and even beyond the boundaries of the body to what Dawkins calls the extended phenotype.

As with the concepts *life* and *intelligence,* it is a mistake to think that diverse theoretical domains should get by with the same notion of phenotype and understanding of "gene 'for.'" My discussion begins with a basic causal analysis of "gene 'for'" that establishes some connection between a gene and a putative phenotypic character. I will describe ways of understanding the terms "phenotype" and "gene 'for'" suitable for the conventional biologist and the investigator into genetic disease. I will then provide definitions of these terms suitable for the unit of selection debate. These definitions will help us face up to the moral gene-centrism threatened by the apparent teleological dependence of phenotype on genotype. To make sense of Dawkins's activity requirement, I propose a set of teleological constraints on the basic causal analysis that maps genes onto phenotype in the way that genic selectionism requires. It does so at the cost of turning Dawkins's picture of the biological world on its head; in the vast majority of cases, it will be much better to say that genotype is teleologically dependent on phenotype. And so, moral gene-centrism will not be an inevitable consequence of taking selected purpose in nature seriously.

How Genes Might Be Active

For genes to be active, they must have systematic causal effects on phenotypes. This is the only way that they could be targeted by the natural selective forces that most immediately affect the phenotype. We should be wary of too simplistic a description of the causal relationship between gene and phenotype. No gene produces anything extracted from a special genomic and extragenomic environment. Some authors have alleged that the context-dependent nature of genic influence represents a decisive blow to attempts to talk about genes for phenotypic characters.[13] This is actually no threat to Dawkins, who builds context dependence into his definition. Central is the idea of change against a constant genomic and environmental background: "When a geneticist speaks of a gene 'for' red eyes in Drosophila . . . [h]e is implicitly saying: there is variation in eye colour in the population; other things being equal, a fly with this gene is more likely to have red eyes than a fly without the gene."[14] This analysis allows Dawkins to justify some of his more outrageous claims. Take the idea that there is a gene for reading.[15] I can read. There are changes that could be made to a gene or genes of mine, in a counterfactual history preserving all the rest of my DNA and environment, that would result in an individual that could not read. This way of understanding the gene/phenotype relationship does not lay itself open to the charge of genetic determinism. The same analysis that shows us that there are genes for reading also points us toward environmental factors for reading—we can agree that a variety of changes could be made to the *environment* of my developing that would have prevented me from reading, or that a genetically identical twin might also result in an individual who could not read.

Too Much Extended Phenotype

The causal analysis described above is the first step on the way toward an account of genic activity. The next question is, at what point on the causal chain leading out from the gene do we find phenotype, or the effects of the gene that constitute its activity? The obvious place would be the boundaries of the individual's body. Dawkins rejects this commonsensical idea, arguing that researchers into adaptation must recognize that genic effects can reach out way beyond the body. He describes the relationship between genes in a beaver and the beaver's dam: "The effect of the dam may be to flood an area thousands of square metres in extent. As to the advantage of the pond from the beaver's [or the beaver's genes'] point of view. . . . By building a dam across the stream the beaver creates a large shoreline which is available for safe and easy foraging without the beaver having to make long and difficult journeys overland."[16] Dawkins holds that "the lake may be regarded as a huge extended phenotype."[17] It does seem to be something that the gene helps produce in order to smooth its passage into the next generation. He adds cases in which the extended phenotype of the genes in the body of one organism may include the body of another organism. The trematode parasite thickens the shell of a host snail. It is speculated that the shell thickening boosts the parasite's fitness by making the snail harder to crush. But this action lowers the host's fitness since the resources used to thicken the snail's shell are subtracted from those available for reproduction.[18]

The above examples are all cases in which the search for properties that are selectively relevant, enabling genes to self-replicate, draws attention away from the boundaries of the individual organism. Dawkins's account of phenotype will need to reflect this.

Before proceeding with Dawkins's account, I will point out that extended phenotype cases should be of special importance to those interested in founding an environmental ethic on selected purposes. As we have seen, environmental ethicists seem faced with a choice between individualistic and holistic approaches to nature. If we can avoid the moral pull of the gene, we have a first clue that concern for individuals' selected biopreferences may lead us to place moral importance on something beyond the individual.

Dawkins appreciates that this story about extended phenotype may lead things to get out of hand.

> Isn't there a sense, it may be asked, in which every possible effect an animal has on the world is an extended phenotype? What about the footprints left in the mud by an oystercatcher. . . . A pigeon's nest is an artefact without a doubt, but in gathering the sticks the bird also changes the appearance of the ground where they have lain. If the nest is called extended phenotype, why shouldn't we so call the bare patch of ground where the sticks used to lie?[19]

How can we stop it from being the case that every part of the world with which an organism and, therefore, its genes comes causally into contact count as extended phenotype? An analysis compelled to cite every genic effect against any background goes far beyond singling out vehicle effects. Clearly, we need to impose some restrictions if we are to arrive at a revisionist notion of phenotype that will bear any explanatory weight.

The Multiple Meanings of "Phenotype" and "Gene 'for' "

In chapter 5 I have urged that diverse interests in the concept of life make a definitional pluralism attractive. The same goes here. Talk of genes and phenotypes can be found in a wide range of fields within the biological sciences. The expressions "gene 'for' " and "phenotype" occupy salient locations within embryology, the study of genetic disease, the science of adaptation, and so on, and it would be surprising if exactly the same notions of "gene 'for' " and "phenotype" could satisfy theoretical needs peculiar to each of these domains. Further, if notions of "gene 'for' " and "phenotype" can differ across domains, we should not be surprised that some accounts depart quite considerably from any commonsense understanding.

I assume that all useful talk about a gene's impact on phenotype will require some manner of causal connection between them. We can make

Dawkins's story about change against a constant background our starting point, representing it as follows:

[CA] A gene *G* is a gene for phenotypic character *P* if and only if a change in *G* against a constant genomic and environmental background produces a change in *P.*

If the conditions outlined in the CA (causal analysis) are met, then we have established a causal connection between *G* and *P.*

One aspect of the CA deserves immediate comment. Dawkins's original proposal includes a probabilistic hedge. He says that a gene is a gene for red eyes in drosophila if "other things being equal, a fly with this gene is more likely to have red eyes than a fly without the gene."[20] In contrast, the CA idealizes by fixing the genomic and extragenomic environments in their entirety. Although this idealization affords great conceptual precision, it is less effective as a practical guide. An experimenter can hope to fix only part of the environment under investigation, leaving the rest variable. Though in practice the probabilistic hedge will do important work, I stick with the conceptually clearer CA.

As it stands, the CA is too undiscriminating either to give a common-sense account of phenotype or to do important theoretical work. Far beyond the oystercatcher's footprints, the CA counts any effect of a gene as phenotype, and this leads to bizarre outcomes along both spatial and temporal dimensions. Phenotypes would stretch out through very long periods. For example, we would seem to have little option but to count the paleontologist's joy as a phenotype of the genes of the apatosaur whose fossil she or he has just uncovered.

I will suggest that the CA might be theoretically useful in some circumstances. However, I think it is best if we treat it as a proto-analysis, an account for which we must find constraints appropriately tailored to our interests.

We must look to a given theoretical domain for appropriate constraints on the CA. There are two domains in which interest in genes and their effects does not often extend beyond the skin. Conventional biologists are primarily concerned with individual organisms, and this concern is not theoretically arbitrary. Individual organisms are causally/functionally integrated units, parts of which tend to enter into regular and predictable relations with other parts, and the skin marks the boundary of that causal/functional integration. For this conventional biologist, the trematode genes and the snail with the thickened shell will belong to different units of analysis. They are potentially independent of one another in a way that parts of individual organisms usually are not.[21] The conventional biologist interested in genes will want to know what effect these genes have on the bodies of individuals, the primary objects of his or her research.

The researcher into genetic disease also tends not to have interests that go beyond the skin. A perhaps operationally useful notion of phenotype for this researcher might even cut the causal chain leading out from the gene earlier than the conventional biologist does. A common strategy attempts to trace certain diseases to deletions or scramblings of DNA; these genetic errors result in an aberrant protein being coded for. Some hope to treat genetic illness at the source. One day, missing bits of DNA might be inserted into the genome, thus enabling the right protein to be produced. Given this goal, a notion of phenotype, or relevant genetic effect, for this type of researcher will help highlight the aberrant or correct protein. For the purposes of this type of research, the assumption is that effects downstream from this genetic defect will be fixed.[22]

Genes and Natural Selection

The person attempting to answer questions about selection brings a quite different set of requirements from those quickly outlined above. There are two ways in which the CA might be constrained so as to illuminate the workings of natural selection.

First, an analysis might guide a broad interest in how a gene's causal effects might make it eligible for selection. Any given causal effect of a gene can incline it to selection in some set of circumstances, so long as those circumstances are capable of being repeated. It is not hard to think up cases in which even the most bizarre causal chains from gene to environmental effect might be a reliably repeated spur to selection. To continue the story about the paleontologist and the apatosaur started above, imagine that sufficient DNA is extracted from the fossil. The dinosaur is then reconstructed by using techniques something like those described in *Jurassic Park*. The resulting creature lives for a while, then dies and is fossilized. After another 150 million years these fossilized remains are discovered. Imagine that this sequence of events is repeated a great number of times. The apatosaur's bones might, under these circumstances, acquire the function to fossilize in an easy-to-find fashion, and the paleontologist's manipulated behavior will be a vehicle effect of the apatosaur's genes as worthy of study as any other adaptation.

Though a very imaginative student of potential fitness might consider the CA unmodified, constraints on the CA will allow a more useful tool for studying fitness. More easily replicable effects of genes will be more likely to help genes into the next generation and are, therefore, more profitably examined by someone trying to catch selection in the act.

In talking about current rather than future adaptation, we must look to past selection. Just as the conventional biologist seeks to isolate causal/functional units, the aim here is for an analysis that will isolate biofunctional units. Such analysis will have the task of pointing toward the knots of phenotype that are, by design, mutually supporting and reinforcing. This historically oriented enterprise is quite different from the one discussed in the above paragraph. For Dawkins, the snail's thickened shell is extended phenotype, or part of the vehicle of trematode genes, because it has been selected for the effect. This is true regardless of the fact that in many present-day cases the shells' extra thickness helps trematode genes not a jot. Nor should we be tempted to count beneficial yet clearly accidental effects as part of a Dawkinsian vehicle. An ornithologist's love for, and therefore disposition to, participate in the preservation of oystercatchers may be inspired largely by the pretty marks that their feet leave in the mud. However, as ornithological affections have played no part in the design and maintenance of the footprints, we cannot describe the prints as adaptations.

Genes for Extended Phenotype

The obvious answer is to build past selection into our account of "genes 'for.' " Phenotype will be conceptually tied to selection. There are two ways in which this might be done. These have reverse implications for the direction of teleological dependence. The first, more Dawkinsian analysis identifies vehicle effects as those that have been selected to enable genes to replicate themselves. The second analysis makes genes teleologically dependent on phenotypes. It identifies phenotype as those effects of genes that genes are selected to produce.

We can represent the first proposal as follows.

[TA1] G is a gene for P if and only if P has been selected for its contribution to the transmission of G.

The TA1 (first teleological analysis) makes phenotypic characters dependent on germ-line DNA. For Dawkins, the TA1 would clearly be a congenial way of telling the story, preserving the teleological primacy of genes. The biocentrist will, however, not look fondly upon the TA1. The replication of DNA becomes the fundamental purpose of any selected biopreference, and to respect biopreferences will be to respect genes.

Another, and I think superior, way of telling the selectional story will invert Dawkins's arrow of teleological dependence:

[TA2] *G* is a gene for *P* if and only if *G* is selected for the production of *P.*

According to our second teleological analysis, the TA2, genes have a phenotype-oriented purpose. It holds that the existence of a given chunk of DNA depends on the existence of a particular phenotypic character.

The TA2, like the TA1, does a good job of matching genes with phenotypic traits that can be considered to be their vehicle effects. Take some of the extended-phenotype cases. There is a trematode gene for thickened snail shells because this gene has been selected for the thickening of snail shells. Were it not for its effect on snail shells, it would, very likely, not exist. The same cannot be said about the oystercatchers' footprints. It is highly unlikely that any gene in the oystercatchers' body has been selected for, or exists now because of, the production of historical footprints like these. This is despite the fact that the gene may play some causal role in the production of these footprints, the footprints may be a typical consequence of the gene, and they may even in some cases benefit the bird. We discover that selfish DNA is not for any phenotype by appreciating that no phenotypic trait depends on this DNA.[23] The very rare but highly beneficial effects discussed above also count as vehicle effects. It is important to remember that the accounts of phenotype and "gene 'for'" that we have customized for the researcher into adaptation will by no means do for all purposes. The conventional biologist wants to count neutral or deleterious effects as phenotype. Neither the TA1 nor the TA2 will do this. This, of course, does not mean that there is no biologically interesting account that groups these features together with selected ones.

If the TA2 can be placed on a par with the TA1, we would have demonstrated a much more teleologically symmetrical relationship between gene and phenotype than Dawkins supposes. We may conclude that it is arbitrary to pick out one level of nature as teleologically primary. In what follows I will go one step further. It is better to say that the function of the gene is to produce the phenotype than it is to say that the function of the phenotype is to generate more genes of the particular type. We should opt for the TA2 over the TA1.

Phenotypic Characters and the Phenotype Kind

First, we must clear up an ambiguity. When we use the word *phenotype,* we can be referring to a kind made up of all phenotypic features, or we can be referring to specific characters such as hearts, lungs, or opposable thumbs. Questions about origins and selectional dependencies of the phenotype kind have quite different answers from questions about origins and selectional dependencies of individual phenotypic characters.

Dawkins's (1990) first chapters present a possible solution to questions about the origin of phenotype as a kind. Dawkins thinks that life began with self-replicating structures that were essentially naked genes. Eventually, there were mutations allowing the construction of novelties such as phenotypes. If this story is right, then there is a sense in which bodies and phenotype exist only to enhance the replication prospects for DNA. The phenotype kind is an adaptation for genes, and we can appeal to the etiological theory to say that the phenotype kind has the biofunction of promoting the survival of the gene kind. By contrast, it is simply false to say that genes came into existence so that more efficient phenotypes could be built. In a physically and biologically consistent world similar to this one, apart from the fact that no mutation allowing the production of phenotype has ever occurred, the self-replicating structures that Dawkins thinks we should call genes would presumably still exist.

Some writers on adaptation or selectional accounts of function insist that we be alert to the work an effect can do in *maintaining* a structure rather than just those effects that bring structures into existence.[24] To take a well-known example, though the breaking up of bamboo stalks does not explain why the panda's thumb came into existence, were it not to be capable of assisting in this bamboo-related job, the thumb would not exist in anything like its current form. This fact permits us to describe the thumb in adaptationist or selected functional language. The same is not the case when we investigate any function that the gene kind might have in relation to the phenotype kind. If we removed all phenotype-causing genes and prevented their return, there would still presumably exist some phenotypeless genes. The existence of the gene kind does not depend on the existence of the phenotype kind—meaning that it has no selected function or purpose in regard to the phenotype kind.

The focus of the student of adaptation is usually finer grained than this. Individual phenotypic features are the objects of study. It is possible for the phenotype kind to depend on the gene kind, as I have just pointed out, and for a specific gene to have a selected function in regard to a particular phenotypic character. This finer-grained approach reveals a high degree of functional interdependence. Were not individual hearts to have boosted gene replication, they would not exist in their current form. If we accept the etiological theorist's analysis of function, this licenses the ascription of a function to boost gene replication. On the other hand, take the specific sequences of nucleotides making up the many genes that produce the proteins necessary for building hearts. In the case of many of these genes, it is highly likely that participation in heart construction has worked to hold their place on the genome. Again, this would license assigning a selected function to genes to aid in building hearts. If this picture is true, the TA1 and the TA2 pick different strands out of this criss-crossing web of biological functions.

If reasoning like the above allows us to place the TA2 on equal footing with the TA1, then we have grounds to reject Dawkins's genic teleological priority thesis. I think that the TA2 is even preferable to the TA1 because statements constructed with the TA2 as template are likely to be more informative than those that take the TA1's form. This is because statements of functional relationships between biological objects tend to be more useful when the causal or developmental chain separating them is shorter rather than longer. In this case, distance between biological objects can be measured in terms of the number of other systems standing between them.

A fairly routine example of the teleological dependence of one kind of biological object on another will illustrate my point. How should we best understand the teleological relationship that exists between certain islet cells on the pancreas and insulin? We know that these cells secrete the hormone required for sugar metabolism. Here, again, counterfactual dependencies seem to go both ways. Islet cells exist, at least in part, because they have played a role in producing insulin. Also insulin exists, at least in part, because it has played some causal role in the production of future organisms with islet cells. Note that whereas the causal chain referred to in the first statement is rather short—it goes from islet cells to insulin—that picked out in the second is comparatively long. A much more immediate function of insulin is the rendering of glucose usable by the body's tissues. This, in turn, has important consequences for the organism's ability to perform a variety of tasks, including reproduction. Reproduction involves the bringing into being of a system that combines genetic and environmental input to create islet cells. So the path from insulin to the manufacture of islet cells is much longer than the path from islet cells to the production of insulin.

Why this prejudice for shorter rather than longer causal chains between objects cited in function statements? The precision of, and information conveyed by, function statements has a tendency to dissipate if we have to trace it through many intermediate systems. To return to insulin and islet cells: we are comparing a statement that assigns to insulin an islet cell–related function with a statement that assigns to islet cells an insulin-related function. There are a number of systems that have the same function as insulin with respect to islet cells. The bloodstream plays a crucial role in allowing the passage of insulin throughout the body. Heart, arteries, veins, and capillaries combine to allow blood to circulate. All these systems add their efforts to those of insulin to get the organism to the point at which it can reproduce and bring into existence new islet cells. The islet cell–related function of insulin will be the same as that of these other biological systems.

If we obey the requirement that the causal chain be short, we can be sure that we are saying something about insulin, blood circulation, or arteries that cannot be as well said about a range of other functional chunks of the body.

Similar considerations should make us prefer the TA2, a statement that assigns a function to DNA in regard to a phenotypic character, over its teleological reversal, the TA1. A heart is causally or developmentally distant from the passing on of its underlying genes. We would expect it to have the same function with regard to this process as many other organs. They help bring the organism to the point at which it can reproduce. At this point, devices with more specific functions—for instance, those involved in meiosis—must be depended upon. By contrast, genes have a quite specific function with regard to certain phenotypic characters. The fact that there are many less intermediate systems allows us to convey more information in pointing to this counterfactual relationship.

There are cases in which statements that take the form of the TA1 seem comparably useful to those that take the TA2's form. The causal chain leading from the sex organs to the passing on of DNA is shorter than that originating from the heart, for instance. There are considerably fewer intervening biological systems. Thus we can assign gene replication–related functions to sex organs in a way that satisfies the causal shortness requirement; in doing so, we say something about them that cannot be said about almost every other phenotypic character. Of course, we can also assign quite specific functions to certain genes with regard to sex organs. Here again, the causal or developmental chain, or number of intervening biological systems, will be comparatively short.

It is important to remember that the TA2 will not do a good job of explaining the terms *phenotype* and *gene "for"* everywhere. The individual organism–focused conventional biologist wants to include the neutral and deleterious effects that the TA2 excludes. Here, different accounts must be chosen. Just as a researcher employing a variant of the CA must remain alert to context dependence for fear of lapsing into genetic determinism, so too the researcher who takes my advice will not forget about the domain dependence of notions of "gene 'for' " and phenotype.

To best reconstruct Dawkins's revisionist phenotype, we have answered his "forbidden question." Many genes will be functional and, therefore, teleologically subservient to phenotypic characters in exactly the same fashion as hearts and eyes.

How does the argument I have presented help us to get around the problem I have described at the beginning of this section? I seek to ground the value of the environment in the biopreferences of individual organisms. The contents of these biopreferences have been described in terms of natural selection. If Dawkins is right, these selected contents are fundamentally gene-directed. The teleological priority of genes will mandate their moral priority. In this section I have resisted the pull of the gene on the selected goals of individual organisms, suggesting instead that it is preferable to talk of phenotypes as teleologically prior to genes. So, we can make selected goals precise in such a way that does not make genes morally salient. At the same time, the analysis I have provided enables us to fully vindicate Dawkins's revisionist conclusions about the extent of the phenotype. It allows us to say that there are trematode genes for thickened snail shells but not that there are oystercatcher genes for footprints. This stands to be good news for an environmental ethic. In isolating teleologically interdependent knots of phenotype that extend beyond the bodies of organisms, an ethic founded on selected biopreferences can hope to show that something beyond the organism is morally important.

SENTIENCE AND GOALS

I have noted in chapter 1 that both intentional and phenomenal states figure prominently in the intuitively attractive moral principles making up the psychological view. I have used the label *rationalist* for those ethics that place importance on intentional states such as plans and projects. *Hedonis-*

tic approaches make morally salient the capacities for pain and pleasure. Though talk of the phenomenal states of nonhumans was once not considered to be scientifically respectable, the work of M. Dawkins (1993), Bernard Rollin (1990), and Singer (1993) has gone a considerable way toward rehabilitating this language. In chapter 2 I have talked about the attempt by Singer and other animal welfarists to transform our intuitive hedonistic thinking into a sentience-based morality. In this view, sentience marks the boundary between those that can be harmed in any morally interesting way and those that cannot. I have shown how concern for sentience might point us toward natural kinds running through the minds of humans and some nonhumans. Our thinking about the moral status of humans includes rationalistic and hedonistic reasons, and our approach to the values of nonhuman individuals should appeal to plausible naturalizations of both intentional and phenomenological states.

Though I have been more interested in states on the rationalist side of the intuitive divide, it is important not to overstate the opposition between the rationalistic and hedonistic approaches to nonhumans. My aim is not to dislodge sentience accounts in favor of the above account. Rather, my opponent is the person who argues for an exclusive focus on sentience.

Note that even exclusive concern for biopreferences and goals would not leave consciousness out in the cold. So long as some story about the causal/functional efficacy of consciousness turns out to be right, we can accord it great derivative importance.[25] Consciousness will open up new varieties of goals to an organism. This greater representational complexity allows an organism to approximate much more closely value-endowing folk psychological paradigms.

I will do more than blandly assert the moral coexistence of sentience-based concern and interest in plausible naturalizations of intentional states. Intentional and phenomenological states ground principles demanding very different and sometimes conflicting behavior from us. My compatibility thesis requires me to detail circumstances in our dealings with natural individuals in which it is more appropriate to focus on sentience, and I then outline the conditions under which we should place more emphasis on biopreferences.

Sentience-Goals

To distinguish them from, and contrast them with, the above-outlined contentful goals, I will call goals generated in respect of sentience, *sen-*

tience-goals. In ways that I will now explain, an organism's sentience-goals are tied tightly to its conscious experience.

The theory of Peter Singer is well known. Singer divides organisms into three broad categories. Least morally considerable are the nonsentient. As we have seen, Singer doesn't think the goods or ills of the nonsentient have any direct place in moral theory. They have no sentience-goals. Moral consideration only comes into play when we look at creatures that are conscious. An organism that is conscious but not self-conscious can entertain only a fairly restricted type of sentience-preference. According to Singer, these creatures "more nearly approximate the picture of receptacles for experiences of pleasure and pain, because their preferences will be of a more immediate sort. They will not have desires that project their images of their own existence into the future. Their conscious states are not internally linked over time."[26] The manner in which the merely conscious experience the world has important consequences for the moral treatment of them. According to Singer, their instant-to-instant existence means that they are not harmed in being painlessly killed.[27] Beings that are self-conscious, or aware of themselves as existing over time, can have more detailed sentience-goals. Singer uses the following thought experiment to demonstrate this:

> If I imagine myself in turn as a self-conscious being and a conscious but not self-conscious being, it is only in the former case that I could have forward-looking desires that extend beyond periods of sleep or temporary unconsciousness, for example a desire to complete my studies, a desire to have children, or simply a desire to go on living, in addition to desires for immediate satisfaction or pleasure, or to get out of painful or distressing situations.[28]

So in order for something to qualify as an object of moral concern, it must be capable of experiencing suffering or pleasure. The kind of respect we owe something with this capacity is determined by what aspects of itself and its environment the being is capable of representing and how these representational abilities feed into positive or negative experiences.

Now that the view is more clearly stated, we can identify differences between considerations grounded in sentience and considerations grounded in biopreferences. First, and more obviously, a focus on biopreferences promises to lead to a morality that is more extensive than Singer's. The class of beings having biopreferences sculpted by natural selection runs from the sentient cow through the barely sentient snake to the certainly nonsentient slug and beyond.

Another key difference lies in the long spatial and temporal reach of many selected biopreferences. According to Singer, the less cognitively sophisticated the organism, the more spatially and temporally short-ranged are its sentience-goals. These animals prefer that their experiences be pleasurable rather than painful. More cognitively complex individuals have sentience-goals extending beyond individualistically characterizable states of an animal's body and brain only when information about these states is capable of triggering positive and negative experiences. Selected biopreferences, by contrast, can be wide regardless of how they feed back into conscious experiences. The vast majority of them can only be satisfied by bringing about a certain set of environmental conditions.

Some writers have commented on the external nature of biofunctional states in a quite different context. Orthodox functionalism has a widely-commented-upon shortcoming in respect of mental meaning. Hilary Putnam's (1979) twin-earth thought experiment is taken by many to show that meaning is essentially broad or environment dependent. Yet the facts that determine an orthodox functional role are specifiable entirely narrowly in terms of relations to other mental states, perceptual states, and behavioral states. They appear not to be environment dependent. Elliot Sober (1990), Sterelny (1990), and McGinn (1989), to name three, have argued that the biofunctional account can make good on this problem. Biological functions are essentially environment dependent. Here is Sober: "A simple fact about functions is that what the function of something is depends on the kind of environment in which it is embedded."[29] McGinn echoes these sentiments: "*The relational proper function of representational mental states coincides with their extrinsically individuated content.* . . . It is not merely that mental states have both content and function; the two are intimately connected, integrally related."[30]

If we apply what we know about the teleological theory of content to selected biopreferences, we can draw analogous conclusions. Having met Fodor's worry about indeterminacy and the challenge that all biopreferences might be gene-targeted, we can see that the teleological theory often identifies as the target of a biopreference a specific relationship between the organism and its environment. The frog's selected biopreference is that it consume a fly, or at least something that is airborne, nutritious, and so on, rather than that it merely experience the sensations that normally accompany successful fly-catching and fly-consuming. Nor is its interest in the fly only as a means of generating pleasurable experiences. This is true regardless of the fact that nature has often designed organisms so that they are sensitive only to the proximal indicators of these goal states. So, selected biopreferences will often have longer spatial reach than sentience-goals.

They also often have a greater temporal reach. By contrast with what will frequently be the case with Singerian sentience-goals, it is highly unlikely that there could be an organism whose selected biopreferences related only to time-slices immediately following the time-slice of the bio-preference. As we have seen, Singer thinks that the failure to project sentience-goals into the future means that many organisms are not harmed in being painlessly killed. The teleological approach to preferences reveals the diachronic interconnectedness of an organism's behavioral projects. The lack of an ability to conceive of oneself as existing over time does not prevent an organism from having a wide range of behavioral projects that require its future existence. For example, it is a selected biopreference of many organisms to find an appropriate mate or mates, reproduce and nurture offspring, and so on.

It might seem that despite their broad nature, for all practical purposes, respecting sentience-goals is really little different from respecting selected biopreferences. We may assign different contents to selected biopreferences and sentience-goals. However, selected biopreferences will tend to be satisfied when sentience-goals are satisfied, and vice versa. After all, natural selection has surely designed animals so that their sentience-goals have tended to be satisfied when broader selected biopreferences are. Yew-Kwang Ng (1995) mounts an argument to this effect. He observes that the physical bases of suffering or pleasure have evolved so that animals capable of complex and variable behavior can select fitness-boosting options: "When a reward and punishment centre in the nervous system had evolved, it helped the survival of individuals in the species concerned if activities (e.g. eating, having sex) conducive to the survival of the gene are rewarded and conditions (injuries, starvation) bad for survival are punished."[31] Ng then goes on to estimate levels of suffering and enjoyment in populations of sentient organisms based on how often they get to perform fitness-enhancing behavior such as mating or eating or fitness-decreasing behavior such as failing to mate and dying. He claims that the Darwinian struggle is bound to produce populations characterized by an excess of suffering over pleasure.

Is Ng's claim about the role of positive or negative sensations as indicators of fitness-enhancing or -reducing behavior true? Kin-selected altruism involves an organism in performing behavior that reduces its own individual fitness in order to boost the fitness of relatives. As a result of this kind of selection, an organism may possess biopreferences that lead to considerable suffering. The beneficiaries of this behavior will often be nonsentient eggs or barely sentient offspring.

Putting this kind of case to one side, there is good reason to think that there will be a far-from-perfect alignment of enjoyment and fitness-enhancing acts. According to Ng, pain and pleasure are indicators of certain categories of behavior. Natural selection can lead to an indicator's being rather poorly correlated with what it is supposed to indicate.

I will now generalize an observation about the relationship between indicators and ecologically important external properties. Earlier in this chapter I have mentioned the indicator theory of content. The indicator theorist holds that the content of a representation is determined by what environmental property it reliably indicates or covaries with.[32] Millikan (1989a) points out that states that are not reliable indicators of a given environmental condition will often get selected to equip the organism to deal with that condition. Take an organism that needs a representation of predators. A representational system that produces very many false positives but very few false negatives will be much more eligible for selection than one that is, on the whole, more accurate but produces more false negatives. An organism guided by the former kind of representational system may often hide or flee when there is no threat, but it is more likely to produce the appropriate behavior when there is real danger. Similarly, natural selection may encourage a link between feelings of pleasure and a certain type of behavior even though most often the behavior satisfies no biologically important goal.

Many animals enjoy sex. The disposition to take pleasure in sex undoubtedly exists now because it boosted the fitness of animal ancestors. However, this pleasure is far from a reliable indicator of fertilization. For example, human females conceal ovulation. In other species, a male's sperm often faces competition from the sperm of other males before it can fertilize any egg. So there will be many times when there is sexual pleasure but no actual boosting of biological fitness.

Thus, concern for sentience-goals may often pull in a different direction from concern for biopreferences.

Captive and Wild Animals

We have seen that biopreferences differ from sentience-goals. Rather than supplanting concern for sentience with a morality grounded in selected biopreferences, I claim that a complete picture of the moral status of animals will take both states into consideration. The two moral pictures will have differing spheres of influence. In the case of animals in their natural habitats, concern for biopreferences comes to the fore. For those removed from their environments and integrated into human projects, on the other hand, sentience-goals will assume more importance.

Let's start with animals in their natural states. In finding an answer to Fodor's indeterminacy objection, we have shown that animals determinately represent aspects of their environments. The correlative claim about biopreferences ties them to goal states in the environment rather than to sensations or pleasure or any other internal marker of success. If we take into account the environment-dependent nature of biopreferences, we must do more than insulate animals from unpleasant experiences.

Robert Nozick's (1974) well-known "experience machine" thought experiment helps to separate the satisfaction of preferences from the achievement of the pleasant phenomenological experiences normally, but not of necessity, connected with them.

> Suppose there were an experience machine that would give you any experience you desired. Superduper neuropsychologists could stimulate your brain so that you would think and feel you were writing a great novel, or making a friend, or reading an interesting book. All the time you would be floating in a tank, with electrodes attached to your brain. Should you plug into the machine for life, preprogramming your life's experiences?[33]

We reject the possibility of existence in the experience machine because central to our conceptions of what it is that makes a life worthwhile are the experiencing of producing behavior that has genuine external world effects that are roughly as we imagine, and being affected by an external world that is approximately as we conceive it.[34] If animals' biopreferences are equally about the external world, a similar conclusion should be drawn. Having just the experiences associated with finding food and a mate, raising offspring, and so on would be a poor substitute for the real thing, whether we are talking about a lizard, a tiger, or a human. Mere satisfaction of sentience-goals will not suffice.

A focus on biopreferences should lead to a dim view of the removal of animals from natural environs and their integration into human projects. Captive conditions can resemble Nozickian experience machines. Here, animals act in ways that often only parody their natural hunting, threat, and mate-attracting behaviors.[35] Familiar behavioral strategies systematically misfire. They suffer harm regardless of the extent to which their predator-free, food-rich surroundings anesthetize them.[36]

There will still be the question of how we should treat animals for whom there is no prospect of escape from captivity or domesticity. In these circumstances, it is often right to discount biopreferences. The situation I am imagining is similar in interesting respects to those in which we deemphasize the projects of humans. Very sick humans may find themselves completely unable to form projects that are fully sensitive to their all-things-considered life plans. Many of the projects of insane humans are entirely inappropriate to their circumstances. Treatment of these people will aim to restore to them the capacity to form and carry out projects sensitive to their environs and life plans. Failing this, or in the interim, unless we have access to earlier formed projects, the correct treatment of these patients must settle for a different goal. We seek to treat them in a way that results in pleasurable rather than painful experiences. To the degree that captive animals are unable to carry out selected biopreferences, we should treat them in a similar fashion.

There is a rather programmatic reason for adopting a view of the value of individuals founded substantially on biopreferences. Many writers have commented on the supposedly irreconcilable conflict between environmentalists and animal welfarists.[37] For example, the saving of ecosystems may sometimes involve the painful elimination of introduced sentient animals. This approach promises to go some way toward showing how animal welfarism might be extended to a broader environmental ethic. An animal's natural environment is the essential background against which its selected biopreferences can be satisfied. In focusing on the endangered wolf, we cannot help focusing on its disappearing habitat. Saving poisoned ecosystems is the only way in which we can meet the demands of the individuals that populate them. In chapter 7 I will examine these broader issues. I look outward from the individual and ask whether a morality grounded in biopreferences makes species and ecosystems morally special.

෮

Species and Ecosystems

THE SHORTCOMINGS OF INDIVIDUALISM

The ethic outlined and defended so far affords protection to a good many individuals excluded by the psychological view. Although cave wetas and black robins are unlikely candidates for most commonsense psychological states, they are straightforward bearers of plausible naturalizations of these states. We, therefore, have grounds to count them as intrinsically valuable, using degrees of plausibility to help us decide how much value to assign. A great deal of work has been done, but we are still some distance from what many would recognize as an adequate environmental ethic. For a start, popular discussion of the plight of the environment is almost never framed in terms of individual organisms. Instead, objects of concern tend to be wholes such as species, ecosystems, or even the entire biosphere. As a consequence, some variety of moral holism is widely held to be an essential feature of environmental ethics.

Now, the moral individualist can certainly say that species, ecosystems, and the biosphere have great worth. They are, after all, collections of morally valuable organisms. What, then, is the real difference between an ethic that values a species as a whole and one that values the individuals that comprise the species? To see what is at stake, we need precise formulations of the theses of environmental value individualism and environmental value holism.

The moral individualist holds that the value of a species or an ecosystem is some function of the values of the individuals that make up these biotic wholes. I won't assume too much about the exact nature of this function. It will be determined, in part, by the larger moral framework in which the value of individuals is lodged. For example, an aggregative consequentialist could derive the value of the whole by summing the values of individuals, each weighted according to its representational sophistication. We would then aim to achieve the best balance of biopreferences satisfied over biopreferences frustrated. Kant himself did not think that any non-human could be an end in itself. However, a Kantian might be sufficiently impressed by the biopreferences of a simple living thing to grant it some measure of end-hood. If so, understanding the value of an environmental whole will be a more difficult matter than for the consequentialist. Simply summing the values of individuals will not appeal. Still, all things being equal, we should prefer that fewer rather than more individual organisms of similar degrees of end-hood be mistreated.

Holists deny that there is any function from the values of constituent individuals to the value of the whole. According to them, reckoning individualistically fails to capture all there is to value in a species, an ecosystem, or a biosphere.

So, when asked how bad is the destruction of a species or an ecosystem, the individualist tries to work out how this event affects individual organisms. The death of the ecosystem is not the same as the deaths of all the individuals; some may survive the death of their ecosystem. Were we to adopt a consequentialist, aggregative function from the values of individuals to the value of the whole, we would subtract the combined values of the survivors from the starting value of the community for a measure of the moral damage done.

This first part of the chapter will outline arguments that expose as inadequate this naive individualistic approach to environmental wholes. Though in many cases the extinction of a species or the destruction of an ecosystem involves the deaths of many individuals—and thus will be abhorred by both individualists and holists—there is a range of situations drawing contrasting diagnoses from individualists and holists. In these situations, concern for the environment appears to demand a holistic approach.

Sober's (1986) argument for the inadequacy of individualism is my starting point. According to him, differences between individualists and species preservationists emerge over the *n + m question*. Sober considers possible threats to the members of the species of blue and sperm whales. For animal welfarists, whom Sober takes as his representative individualists, what matters is the level of sentience of blue and sperm whales. By contrast, for the species preservationist,

> . . . a holistic property—membership in an endangered species—makes
> all the difference in the world: a world with n sperm and m blue whales

is far better than a world with $n+m$ sperm whales and 0 blue whales. Here we have a stark contrast between an ethic in which it is the life situation of individuals that matters, and an ethic in which the stability and diversity of populations of individuals are what matter.[1]

So, individualistic theories are stated in terms of the properties an organism possesses regardless of its membership in any group. By contrast, for species preservationists, an organism is more or less important depending on facts about the group it belongs to. The key question is whether or not its species is endangered.

We do not have to look hard for real-life situations that bring the tension between holists and individualists to the surface. To save the native New Zealand black stilt, the feral cats, rats, and ferrets of the Mackenzie Basin in Canterbury may have to be trapped and killed. Individualists who emphasize the significance of sentience have a problem. Cats, rats, and ferrets almost certainly have a much richer array of sentience-preferences than do stilts. This problem is exacerbated when we consider attempts to save the weta, a nonsentient insect, from mammalian predators.

Furthermore, saving the species of black stilts is not the same as protecting all of the current individuals from suffering. Removing each stilt from its hazard-filled natural habitat and providing it with a safe, food-rich environment in which it could not mate would arguably benefit each individual while exterminating the species. On the other hand, the moral cost of pain that individuals might suffer within a successful captive-breeding program would seem difficult for the individualist to justify.

Individualists less preoccupied with sentience also find the going difficult. I have discussed Taylor's life ethic in chapter 4. His focus on teleological centers of life does not allow the needs of the cognitively complex cat to trump those of the cognitively simple weta. However, Taylor gives no straightforward means to discriminate among the various organisms affected by any course of action when species preservationist issues arise. For him, we have a choice between letting the good of individual stilts be destroyed or actively interfering with the presumably equally valuable cats, rats, and ferrets.

I have suggested that certain varieties of individualism pull in a different direction from concern for species. There is a similar tension between these varieties of individualism and ecosystemic value. Someone interested in sentience will find rather few morally valuable individuals in most ecosystems. It may be relatively easy to single out the rather few beavers, deer, or possums and protect their interests. However, such actions will have consequences that are highly inimical to ecosystemic flourishing, whichever way we interpret it. According to J. Baird Callicott, protecting the ecosystems requires

> . . . trapping or otherwise removing beavers (to all appearances very sensitive and intelligent animals) and their dams to eliminate siltation in an otherwise free-flowing and clear-running stream for the sake of the complex community of insects, native fish, herons, osprey, and other avian predators of aquatic life which on the anthropocentric scale of consciousness are "lower" life forms than the beaver.[2]

The above cases are presented to show the error in thinking individualistically about species and ecosystems. Preservationist concerns seem equally out of place when dealing with questions about individual value. Singer warns against speciesism in assessing whether a given sentient being is harmed. The species membership of a chimpanzee makes no difference to the pain it feels in a medical-research laboratory. According to the variety of individualism proposed by Varner (1990, 1998) and quickly described in chapter 4, a butterfly possesses the same good by virtue of the biological functions of its parts, whether or not it belongs to the last breeding pair of its species.

Sober despairs of any resolution of the "deep theoretical divide" between individualism and species preservationism. He is far from alone. As part of his wide-ranging discussion of species preservationist ethics, Bryan Norton proposes for reasons similar to the above that "it would appear that all theories that generate the intrinsic value of species from the intrinsic value placed on individuals can be eliminated from serious consideration as a support for species preservation."[3]

ENVIRONMENTAL VALUE HOLISM

A theory that fails to account adequately for species and ecosystems simply is not an environmental ethic. This seemingly incontrovertible claim, combined with the apparent invisibility of species or ecosystems to moral individualists, has prompted some to argue for a distinct holistic brand of value. Before we consider arguments for environmental value holism, I'll quickly canvass the views about how individualistic and holistic value will interact.

More moderate commentators think that any adequate environmental ethic will incorporate both types of value. Lawrence Johnson (1993) recognizes morally considerable interests in both individual organisms and environmental wholes. Rolston (1988), too, juxtaposes stories about individualistic and holistic value.

Things may not be so straightforward as this compatibilist view seems to suggest. I've just described situations in which individualistic value and holistic value appear to conflict. For a commitment to holistic value to be able to protect endangered species, we must be careful to leave aside some distinctive and protected role for it. Here is where the advocate of cohabitation runs into trouble. In chapter 4 I have noted that attempts to supplement biocentric value with human-centered value leave open a mere lip-service acknowledgment of the former. There is a similar risk here. Certainly, there are well-worn pathways leading from recognition of the value of certain individuals through to action. Comparatively novel holistic value needs not only to be acknowledged as genuine but also to be sufficiently morally motivated for it not to be a practical irrelevancy.

Some ethicists struggle hard to carve out a protected area for holistic value. Here is Callicott on how the land ethic resolves the tension between concern for natural individuals and biotic wholes: "The land ethic not only provides moral considerability for the biotic community per se, but ethical concern for its members is preempted by concern for the preservation of the integrity, stability, and beauty of the biotic community. The land ethic, thus, not only has a holistic aspect; it is holistic with a vengeance."[4]

Rolston aims at a similar conclusion by tilting his account of individual value in a holistic direction. For him, the idea that individual value survives extraction from the relevant biotic whole is an illusion.

> Intrinsic values cannot be supposed to possess and retain their value without reference to anything else. We may say that a tiger has intrinsic value for what it is in itself, but if we were to transport a tiger to the moon, would this intrinsic value remain? No, because the tiger does not have any intrinsic value that it can, all by itself, take to the moon; the tiger is what it is where it is, in the jungle, and that means that the question of intrinsic value has to stay located in its appropriate place.[5]

The claims that holistic value preempts individualistic value or that individuals only posess value by virtue of belonging to biotic wholes do give holistic value protected status within our moral thinking. In this chapter I argue that as holistic value comes into focus we lose sight of the value of individuals and that this change in focus has bad implications for nature.

I will urge that we have no need for distinct holistic value to understand why species and ecosystems are morally important. If we can find the right answer to the question about the values of individuals that make up a biotic whole, we will have the right answer to the question about the value of the whole. Biotic wholes will be fully accommodated within an individualistic framework.

Distinguishing Ethical and Explanatory Holisms

We need to get clearer about the content of the moral holist's claim. I begin by anticipating a potential source of confusion. In what follows I will separate the debates between *explanatory holists* and *atomists* in the context of *scientific explanation,* from the debates between *holists* and *individualists* in *ethical contexts.* These distinct debates are often jumbled together with arguments for scientific holism wrongly taken to support a holism about value.[6]

It is a good idea to start by considering both distinctions as applied to ecosystems. Explanatory atomists think that species and ecosystems can be fully explained by theory about their parts. An atomistic theory of the ecosystem will talk, in the first instance, about only the behavior of individual organisms and various abiotic influences. The explanatory holist counters that any such description is bound to leave important information out. Ecological laws or generalizations will make indispensable a reference to groups or collections. So, in knowing all there is to know about individuals, we are not guaranteed to know all there is to know about ecosystems.

James Lovelock often uses arguments for an explanatory holism about the biosphere to support an ethical holism of it. According to traditional models, the evolution of living things is best viewed as having been directed by changing conditions on the earth. The adaptations of early organisms are held to have tracked, among other things, global cooling and the oxygenation of the atmosphere. By pointing to the many ways in which an organism, in exchanging nutrients for metabolic by-products, modifies its environment, Lovelock seeks to reverse this explanatory arrow. According to him, rather than meekly adapting to changing conditions, life has engineered conditions on earth to meet its needs. Levels of the various atmospheric gases originally arranged so as to permit the evolution of more complex life forms are now regulated to allow their perpetuation. Lovelock says Gaia is "a system that has emerged from the reciprocal evolution of organisms and their environment over the eons of life on Earth. In this system, the self-regulation of climate and chemical composition are entirely automatic. Self-regulation emerges as the system evolves."[7]

The explanatory holistic nature of the Gaia hypothesis is apparent. Focusing on only one organism or restricted biotic community may give a picture of the inputs from, and outputs relevant to, its immediate neighbors. We fail, however, to understand its larger role.

Lovelock thinks that the Gaia hypothesis leads to important ethical insights. Indeed, anthropocentric conclusions follow with little further argument. An atomist may tell us how changes to atmospheric carbon levels or lake trout levels will affect some immediately adjacent systems. Such an account may mislead us into thinking we can rearrange Gaia's parts as we please. Unless we think holistically, or so Lovelock's reasoning goes, we miss the dangerous delayed effects of these changes on distant parts of Gaia and on ourselves.

Lovelock, however, is not satisfied with exclusively anthropocentric conclusions. He hopes that the holistic interdependence of Gaia's parts will counteract the tendency of human welfare to monopolize ethical concern.

The bad news for him is that the distinction between ethical holism and ethical atomism differs from and crosscuts the distinction between explanatory holism and explanatory atomism. When subjecting an object to moral evaluation, we ask how some preferred set of ethical principles applies to the object as a whole, or to its parts taken separately. This is quite different from asking what role a theory about the object's parts should play in explaining the overall behavior of the object.

Explanatory atomism about a given entity can perfectly well coexist with an ethical holism about it. Imagine someone who takes a radically reductionist, atomistic approach to the causes of human behavior. This person claims that beliefs and desires can be completely understood as complexes of neuron firings, and that once we have the full neurobiological story, commonsense intentional psychology will have nothing to add. Regardless of this conclusion, we might still say that, in respect of neurons, value attaches holistically rather than individualistically. A reductionist might well accept the psychological view about intrinsic value described in chapter 1. Although thinking that all of a person's behavior can be traced back to patterns of neuron firings, this reductionist can deny that the value of the person is some function of the values of individual neurons. A single neuron will have no value independent of its contribution to the human psychological whole.

To illustrate, we can perform the following value-subdivision thought experiment suggested by Rolston's tiger-on-the-moon story. Imagine we were to remove a person's neurons, one by one, implanting each in a separate recipient. These neurons would not be replaced in the donor. Further, suppose that we carry out the multiple operations in such a way that preserves the physical integrity and biofunctioning of each neuron. The value we attach to the individual would not track, in subdivided fashion, the destination of each neuron. Rather, the value of the original person would be utterly destroyed.

Take the reverse case, in which an ethical individualism will accompany an explanatory holism. Someone who takes a moral interest in the fortunes of the nation and holds that such phenomena as cultural impoverishment and declining national optimism are essential to this ethical inquiry might also accept a holistic explanation of these phenomena. Such a person denies that these phenomena can be fully understood in terms of the behavior and intentional states of individuals.

This same person may assert an ethical individualism with respect to these national properties. A high degree of national optimism is to be sought only because it covaries with a high aggregate of individual well-being.

This time our value-subdivision thought experiment will give a reverse result. Imagine that a number of citizens decide to emigrate, settling in a wide variety of foreign lands. Our cultural critic will deny that the group-level properties of national pessimism and cultural impoverishment can, in any way, be mapped onto the emigrants and followed to their separate destinations. This critic may accept, however, that any value inherent in the nation is not destroyed but instead tracks the destinations of relocating citizens.

The evident gap between explanatory and ethical holisms means that Lovelock has more work to do before he can draw any conclusions about nonanthropocentric value. On its own, an explanatory holism about Gaia's parts does not support a Gaian nonanthropocentric value holism. We can perfectly well agree that the biosphere cannot be fully explained by reference to the activities of constituent individuals, but also maintain that individuals alone are the bearers of intrinsic value.

Tailoring Moral Principles to Environmental Wholes

How, then, do we go about arguing that a species or an ecosystem is valuable holistically? Many have followed Aldo Leopold's (1949) lead in staking the value of species and ecosystems on their "integrity, stability, and beauty." These terms—*integrity, stability,* and *beauty*—join others such as *diversity* and *complexity* in descriptions of the value that attaches to wholes and is supposed to be resistant to value-subdivision. Holists say that in sending each member of the last population of black robins to separate zoos we do not subdivide any integrity-value or stability-value. Rather, we destroy it.

In what follows I will argue that commitment to the non-subdividable values of species and ecosystems fails to offer moral guidance in respect of them.

The existence of more than one naturalistically acceptable account of species leads to practical dilemmas for the holistic species preservationist ethic. When concern for "species" naturalistically interpreted one way conflicts with concern for "species" naturalistically interpreted in another way, we need to make a choice. Now, it is not out of the question that we might locate in the concept *species* the resources to guide this choice. In chapter 5 I have argued that a plurality of naturalistically acceptable accounts of "life" represent no barrier to the moral importance of life precisely because we can make a significantly strong case for the moral importance of one notion. The bad news is that current holistic species ethics do not give moral notions sufficient purchase on natural properties and so provide no assistance.

When we turn to ecosystems, we find similar problems. Here, the principal worry is not about a plurality of naturalistically acceptable, but conflicting, accounts of ecosystems but is instead about the properties that endow ecosystems with value. It turns out that these properties cannot be promoted or respected without prospect of conflict. Again, the ethics under investigation offer no guidance about how these conflicts are to be resolved.

Which Species?

First to species, beginning with quick outlines of the holistic theories of species value due to Rolston and Johnson. According to Rolston, "the species line is a vital living system, the whole, of which individual organ-

isms are the essential parts. The species too has its integrity, its individual-
ity, its right to life (if we must use the rhetoric of rights); and it is more
important to protect this vitality than to protect individual integrity." Rolston draws out further implications of this species "right to life": "The
right to life, biologically speaking, is an adaptive fit that is right for life,
that survives over millennia. This idea generates at least a presumption
that species in a niche are good right where they are, and therefore that it is
right for humans to let them be, to let them evolve."[8]

Johnson claims that we have a duty to respect a being's interests. Like
Taylor, Johnson rejects the idea that something requires certain types of
mental states in order to have interests. All that is required for a thing to
figure in moral theorizing is that it have a well-being. Among things with
well-beings are species and ecosystems. Johnson says, "Both species and
individuals need to fulfil their nature in an appropriate environment. A
species has an interest in continuing in equilibrium with its environment,
in fulfilling its nature as a species, and in its individual species members."[9]
Species must do more that just survive. Relict zoo populations cannot satisfy many of a species' interests. "It is a matter of what survives. Relating to
a particular sort of environment is part of the self-identity of a species—
and the pressure from the environment helps a species to maintain its self-
identity. What does a species profit if it gains survival and loses its soul?"[10]

This is all well and good, but if we are to arrive at a workable ethic, we
need some way of identifying the things that are supposed to be valued. It
cannot help us to learn that species are morally important only to discover
that we have no way of saying which parts of nature we are to protect.

This is not to say that there has been no controversy over what to count
as an individual organism. Biologists have challenged the identification of
organisms with the reasonably causally integrated and spatially isolated
objects favored by common sense. Though common sense finds many individuals in a field of dandelions, David Janzen (1977) urges that the genetically identical dandelions jointly make up one individual. He points out
that asexual reproduction is not so very different from the mitotic cell division that occurs when any organism grows. In both cases, we have the split-
ting of a cell into two genetically identical offspring cells. Just as it would be
wrong to think of my arm as a different individual from my body, so too we
should avoid drawing lines through a field of dandelions. Dawkins
responds to Janzen by seeking to show that an interest in the evolutionary
significance of the organism does allow the field of dandelions to be subdivided into individuals in a way that largely agrees with common sense.[11]

What should the environmental ethicist make of this exchange? It may seem to threaten the moral position outlined in chapters 4 and 5. Biologists seem to have more than one way of singling out organisms, thereby generating different stories about organisms' interests. So there can be no one individual organism–targeted ethic. The individual-organism ethicist has a swift response. If we accept the psychological view about intrinsic value, we can break any deadlock. The theory about folk psychological states and their plausible naturalizations gives us the tools to decide what manner of organism matters in any given situation. Though Janzen may count me and my clone as one individual, we have our own separate plans and projects. We are, therefore, valuable separately.

Species are the lowest units of classification in conventional biological taxonomy. We would hope, therefore, that they are relatively straightforwardly identifiable. Unfortunately, there is more than one proposal for how we go about determining which organisms constitute a species. Sober (1993) describes three species concepts. Ernst Mayr has proposed a *biological species concept* that makes the idea of reproductive isolation central. According to him, a species is a collection of actually or potentially interbreeding organisms.[12] Pheneticists attempt to use morphological similarity as a basis for grouping organisms. To determine which organisms make up a species, we measure a large number of features and arrive at an overall measure of similarity. Sober labels this the *phenetic species concept.* To these first two we can add the *ecological species concept* of Van Valen (1976). This notion groups together organisms that face the same adaptive problems and evolve together.

Which notion of species is morally important? Perhaps all of them. Each of the three concepts listed by Sober conceivably picks out a thing that, in Johnson's terms, "has an interest in continuing in equilibrium with its environment, in fulfilling its nature as a species, and in its individual species members."[13] To use Rolston's language, each of biological, phenetic, and ecological species seems a candidate for "a vital living system, the whole, of which individual organisms are the essential parts."

Unfortunately, real conservation choices rule out commitment to all at once. Consider again the endangered New Zealand black stilt, reduced to around eighty individuals.[14] Matters become somewhat complicated when we ask for detail on threats to the stilt. Added to predatory rats and feral cats on our list of threats are the seemingly misguided reproductive efforts of the black stilt itself. The pied stilt is a much more plentiful Australian relative of the black stilt. Differences in coloration, behavior, and relative geographical isolation lead conservationists to say that these are two different species of stilt. Yet black and pied stilts will interbreed, and this is precisely what they are doing. Conservationists worry that the black stilts are literally breeding themselves out of existence.[15]

Or are they? If we follow the biological species concept, then we should group black and pied stilts in the same species. Commonsense views about species boundaries will be exposed as simply mistaken, and the black stilt will be removed from the endangered list.

Morally enshrining different species concepts leads to the reverse judgment. For example, behavioral and morphological differences feed into calculations made by the pheneticist that would put pied and black stilts into separate species. Saving the species will require constructing a reproductive barrier where none exists in nature.

This problem would not be serious if we could decide exactly why species are supposed to be valuable. Such an account would presumably fix on a property possessed by certain collections of organisms falling under the name *species* but not others. If it turns out that the property is maximized or preserved by keeping the pied and black stilt separate, then perhaps the phenetic species concept rather than the biological species concept will be the important one. The bad news is that none of the nonanthropocentric holistic explanations of species value mentioned above says enough to point us in the right direction.

Which Property of Ecosystems?

How will properties such as stability, diversity, integrity, and beauty guide us with respect to real ecosystems? To get a grip on this question, we need a basic description of ecosystems. The following first-year ecology-text account will serve as backdrop for an investigation of any holistic moral goods.

An ecosystem combines a biotic community with its physical environment. Biotic communities are made up of populations of organisms interacting with one another in specific ways. Some of the most important community relations are trophic, or feeding, relations. Producers, or autotrophs, such as plants, introduce energy into a community by synthesizing organic compounds from inorganic raw materials. On the next step up are the herbivores that eat the plants and are in turn consumed by the predators. Many of these predators are then prey for other organisms. Decomposers, such as fungi and bacteria, break down the dead remains of other community members, rendering them suitable to be taken up by producers.

Direct trophic interactions among species at different levels have a regulatory effect on population numbers. An unexpected boom in the numbers of wolves will tend to be followed by a bust in the population of deer. As prey levels drop, predator numbers are bound to follow. In the next stage of the cycle, prey levels recover.

Other regulatory influences become apparent as we tighten our focus from trophic levels to individual species. It is a basic principle of ecology that each species can be mapped onto a niche within an ecosystem. In specifying a niche, we say exactly how its occupants use their biotic and abiotic environment, taking care to include information about the manner of prey and where it is to be found. Niches in a species-rich ecosystem may be very densely packed. Seemingly negligible differences in the beak lengths of birds belonging to two species of finch can affect the average sizes of nut seeds consumed and, thereby, distinguish two distinct niches.[16] If an established species and an immigrant are too similar in feeding strategies and food preferences to fill the same niche, competition may result either in one species' being driven out of the ecosystem or in the less efficient competitor's modifying its feeding strategies or prey preferences.

With this brief description in hand, we can understand some moral claims about ecosystems. Many ecologists think that shielded from human interference, an ecosystem tends toward stability and the maintenance of its integrity and diversity. The identities of species within an ecosystem and their relative abundance will be more or less locked into place by the pattern of biotic relationships. Though this equilibrium may occasionally be shaken, the ecosystem will seek out its original balance.[17]

Those same ethical holists who emphasize the apparent tendency of species to maintain their integrity are very much impressed by this apparent capacity of an ecosystem to self-regulate. Johnson's argument for the moral considerability of ecosystems proceeds in a similar way to his argument about species. He thinks that an ecosystem has well-being interests: "We may think of an ecosystem as an ongoing process taking place through a complex system of interrelationships between organisms, and between organisms and their nonliving environment. The organisms change somewhat, but there is a continuity to the ecosystem, and a center of homeostasis around which the states of the ecosystem fluctuate, which defines its self identity."[18] Johnson describes how the well-being interests of ecosystems can be harmed: "An ecosystem can suffer stress and be impaired. It can be degraded to lower levels of stability and interconnected complexity. It can have its self-identity ruptured."[19]

There are problems for this view about ecosystemic interests. When attributes of ecosystems favored by the holist—stability, diversity, and integrity—are stated in sufficient detail to be located in real ecosystems, they threaten to conflict. If they cannot be jointly promoted, we must choose among them. This choice will need to be adequately morally motivated, and I suggest that the value holist is hard pressed to do this.

Consider, first, the view of some ecologists that enhanced stability and high levels of diversity are often incompatible properties of ecosystems. Nature is certainly not all harmony and balance. Storms, fires, and human intervention allow new species to enter a community. A focus on these disturbances has given rise to a new approach to ecosystems. P. White and S. Pickett, advocates of this new approach, define a disturbance as "any relatively discrete event in time that disrupts ecosystem, community, or population structure and changes resources, substrate availability, or the physical environment."[20]

All ecologists agree that disturbances occur. For traditional equilibrium theorists, disturbances are exceptions to the rule of balance and harmony. Disturbance theorists allow that the forces described by equilibrium ecologists have the potential to guide an ecosystem toward equilibrium. However, they claim that such a state is almost never reached. According to Seth Reice, "the normal state of communities and ecosystems is to be recovering from the last disturbance. Natural systems are so frequently disturbed that equilibrium is rarely achieved."[21] Disturbance theorists not only claim that such events are frequent but also accord them an important ecosystemic job: "Nonequilibrium theories attribute the high diversity of species and the coexistence of similar species that we observe to processes of disturbance and recruitment."[22] How could disturbance promote diversity? Although the immediate effect of a disturbance is always to reduce the number of individuals within an ecosystem, fires, floods, and storms cull selectively. To see this, we need to examine how constituent species are affected. Reice notes of the redistribution of individuals following a disturbance that species *richness,* or the number of species in a community, changes only rarely. Flooding and fire almost never remove entire species. Instead, species *evenness* is modified. Evenness is a measure of the numbers of individuals belonging to a species relative to the numbers of individuals belonging to other species.

White and Pickett (1985) note that disturbance-initiated changes in evenness may be all that permit the survival of species that in a stable ecosystem would be driven out by more efficient competitors. Disturbances also allow the introduction of new species, by promoting patchiness. A storm or fire may ravage some parts while leaving other parts untouched. Although life goes on much as before in the untouched regions, new species can colonize the disturbed patches. Reice summarizes the disturbance theorist's view about the relationship between disturbance and biodiversity: "Under a disturbance regime that is intermediate in frequency, magnitude and intensity, some resident species persist in the system along with colonizing species, which exploit the disturbed areas. Thus intermediate disturbance leads to maximum species richness."[23]

The disturbance view threatens to tear apart the cherished goals of ecosystemic equilibrium and high species diversity. Ecosystems will be stable only in circumstances so ideal as to be almost never realized. Further, this lack of stability may be essential to high species diversity.

What implications are there for moral theory if the nonequilibrium view about ecosystems turns out to be true? Were only one of the properties of stability and diversity to be supported by sufficient moral reasons, disturbance theory would pose no problems whatsoever. Practical difficulties would flow from an adequate ethical motivation of *both* properties, though conflicts between valuable properties are certainly not new to ethical theories. I have taken our hesitation when faced with disturbance scenarios as evidence for a third alternative. It demonstrates the impoverished moral support for the nonanthropocentric value of both stability and diversity. Although it is not hard to think up anthropocentric reasons for preferring diversity to stability, or vice versa, nonanthropocentric reasons seem hard to come by. certainly holists' stances about why ecosystems matter morally do not point us toward them.

It might seem, given the discussion of chapter 3, that I am missing a rather obvious way of morally motivating environmental wholes. Individual living things are intrinsically valuable because they contain plausible naturalizations of folk psychological notions. Why should we not also find plausible naturalizations of the same or different psychological notions in species and ecosystems considered as wholes? I have cast doubt on the moral utility of one plausible naturalization at the end of chapter 3: Schull's discovery of intelligence in species.

Most value holists will be hostile to any attempt to view environmental wholes in folk psychological terms.[24] Environmental ethicists with holistic sympathies tend to demand that we defend biotic communities on their own terms, rather than just as third- or forth-best deservers of the language we use to describe ourselves.[25] According to these people, the use of psychological terminology will be yet another symptom of human arrogance.

So far I have argued that it is difficult to defend species or ecosystem value in terms conceptually alien to the psychological view. I will now give reason for thinking that however we choose to motivate holistic value it cannot play a dominant role in our moral approach to nature. Ethical holism is necessarily insensitive to the intuitively huge value differences between different species or between different ecosystems.

An ocean pelagic ecosystem is a grand thing. Krill support a variety of organisms, ranging from whales to squid, seals and birds. Careful study reveals a complex pattern of direct trophic and competitive relationships. With such dramatic niche occupants, it is easy to see that such a thing could be very valuable. However, any determination to ignore the identities of individuals leads to difficulties.

Take a community that contains members interconnected by a similar pattern of relationships to that which we find in the oceans. The mammalian gut contains a complex array of organisms, some breaking down food consumed by the mammal, others feeding on the by-products of the mammal's digestion. These organisms are specially adapted to live in warm, anaerobic, and often highly acidic conditions. Community members include symbionts that benefit through assisting the host by breaking down its food. Other members are parasites such as nematodes or tapeworms. We find in the mammal's gut the same regulatory mechanisms that we would normally find in the pelagic ecosystem. The community is prone to disturbance by antibiotics or salmonella bacteria.

If we have access only to the principles and generalizations of ecology, we find a great many similarities between the two ecosystems. Nevertheless, I would suggest that a pelagic ecosystem has a value that vastly outweighs that of the community occupying a cow's gut. How can we account for this difference? The obvious way would be to consider the separate values of the individuals constituting the ecosystems in question. The holist may agree that we must switch focus in order to see what makes the pelagic ecosystem so much more valuable than the mammalian-gut ecosystem. Perhaps this can be done without telescoping down to the level of the individual, but instead by pulling back to expose the roles the various ecosystems play in the biosphere. In this way, we appreciate their relative importance. While the destruction of a gut ecosystem does not have widespread ramifications, the elimination of pelagic ecosystems stands to impair the proper functioning of the biosphere as a whole.

This response is only partially successful. There is much debate as to what impact the loss of a given ecosystem has on the larger biosphere. It is possible that some version of the disturbance theory holds true of the biosphere, meaning that survival of certain significant ecosystems and of the biosphere as a whole will depend on the frequent generation of ecosystem-size patches. These empirical questions are difficult to answer. We can make some judgments about ecosystem value independent of them, however. Imagining that neither the removal of the pelagic nor the gut ecosystems has any broader biospheran 'impact, it still seems wrong to say that they are equally valuable.

To see what it is that makes the destruction of a biotic whole the tragedy that it is, we need to look at the fates of constituent individuals. Almost any individualist can see why the pelagic community is of so much greater value than the mammalian-gut community. The pelagic community has individual members each of whose value substantially outweighs that of any occupant of a mammal's gut. When we plug the values of these individuals into our preferred function from individualistic to holistic value, we are bound to find that the pelagic-whole counts for more than the gut-whole. Will this individualistic approach mean that species as wholes and ecosystems as wholes drop out of the picture altogether? Not at all. Focusing on the goals of individuals does, in fact, make species and ecosystems morally salient. Of course, we must take the right individualistic story as our starting point.

INDIVIDUALISTIC ETHICS OF SPECIES
AND ECOSYSTEMS

At the beginning of this chapter, I have pointed to the shortcomings of some individualistic approaches to species and ecosystems. I contend that such ethics have not subjected the appropriate properties of individual organisms to scrutiny. By focusing on the biopreferences of individual organisms, we can draw the right conclusions about species. Although we must qualify the naive individualist's function from individual worth to the worth of a whole, these qualifications can be motivated in purely individualistic terms. We should preserve species and ecosystems because in preserving them, we best meet morality's requirement to respect or maximize the satisfaction of individual biopreferences.

In chapter 5 I have relied on a distinction between self and other-directed biopreferences to show that internal organs do not qualify for value independent of the individuals housing them because their biopreferences target the goods of the individual. Most well-known individualists construe the interests of organisms as largely self-regarding. For Singer, organisms prefer that they go on living, not endure pain, and so on. Regan thinks that self-consciousness puts a barrier of rights around the individual, protecting it from harm. According to Taylor, respect for an organism is respect for the life processes of that organism.

The success of the individualistic species ethic will hinge on both other-directed and *other-requiring* biopreferences. As I have discussed in chapter 5, other-directed biopreferences are targeted at others' goods. Other-requiring biopreferences, while not targeted at the goods of other individuals, can be satisfied only if significant other biopreferences of specific other individuals are satisfied. They require other organisms to achieve important and central goods.

One view about how natural selection works might allow us to make a quick jump from the significance of representational goals to a species preservationism. According to this view, evolution by natural selection is essentially the story of the struggle of species against species.[26] The goal of all adaptations of individual organisms is to boost the survival chances of their species. If this is the case, then we should expect the same of psychological adaptations. Their other-regardingness will be species-oriented. Natural selection will build simple organisms with the very important biopreference of preserving the species. So, respect for species will follow from respect for individual biopreferences.

This view of evolution has been discredited. Pointing to widespread intragroup and intraspecies competition, Williams has argued against selection for entities larger than the individual.[27] Individuals that boost their own reproductive ends at the expense of their species are most often fitter than their altruistic conspecifics, and this means that selection in favor of group or species adaptations is usually overwhelmed by selection targeted at the individual or gene. An organism that indiscriminately aids others in the group will be outcompeted by a co-group member that targets its efforts more specifically at relatives. It seems we need to find a less direct route from selectionally endowed goals to species preservationism. Individual biopreferences that will be favored by individual or gene-targeted selection will have to be found.

We've already noted that some organisms have central other-directed goals. The pied stilt, close Australian relative of the black stilt, exhibits risky "broken wing" behavior when its nest is threatened by a predator. The aim of this feigned injury is to make the stilt, rather than its eggs or young, the main target of the predator. This behavior is not to be accounted for in old-fashioned group or species-selectionist terms. Rather, kin-selection is the explanation. The stilt's behavior has evolved because stilts that risked their lives tended to leave more descendants than those that were more cautious. Sterile worker bees care for the genetically identical offspring of the queen. A bee will die in front of an invader to the hive in order to slow its progress. Bees that "commit suicide" in this way make it more likely that there will be later generations of genetically identical bees. This is merely to scratch the surface of self-sacrificing, but inclusive fitness boosting, behavior in nature.[28]

In preventing a worker bee from killing itself by stinging an invader to the hive or stopping a pied stilt from performing its broken-wing behavior in front of nest-bound predators, we act in ways that seem protective of the individual. The individualists discussed earlier might approve; the pied stilt's broken-wing behavior risks the mother to save the less-sentient eggs or offspring. I prefer the reverse analysis. The causal salience of exclusively other-regarding goals over self-regarding biopreferences in these situations makes the intervention that might be preferred by Singer and Regan inappropriate. In saving their lives, we don't treat the bee and the stilt in a way that respects their biopreferences.

These cases of biological altruism are interesting examples of a causally salient environmental-directed goal that is essentially other-organism directed. We have not yet gone far enough to make species morally special. The arguments of Dawkins and Williams cast doubt on the existence of individualistically selected altruistic behavior directed at preserving the species as a whole. As we saw, selection for such behavior will have been swamped by individual or gene-centered selection. There are too many situations in which an individual will profit either by doing a conspecific in or by refusing to come to its aid.

I suggest that we can find other goals of individual organisms to help support a robust species-preservationist ethic.

Sexually reproducing organisms need to find conspecifics with which to mate in order to pass on their genes. Representations in these organisms will be shaped so as to help them achieve this task. Internal structures will be linked with behavior and perception so as to produce certain mate-ward movement, or mate-attracting displays in response to specific perceptual clues. This other-regarding (rather than other-directed) biopreference may be a candidate for the most important goal in sexually reproducing organisms. An ethic grounded in goals will lead us to say that failure to satisfy this biopreference may, by its own lights, be the worst thing that can happen to an organism.

Wilson's description of the Bachman's warbler offers a compelling example of an individual suffering from this kind of harm. Wilson gives the account of a bird-watcher sighting what must have been one of the last Bachman's warblers, a male, singing to attract a mate. During the two hours for which it sang nonstop, no female appeared. Wilson continues: "The male returned to the same spot the next two springs. No female ever joined him. The extraordinary exertions of the Bachman's male were a sign that he was in prime breeding condition, but he was destined to go undiscovered by any female of the same species."[29] By warbler standards, this individual may have led a relatively painless and long life. According to the analysis I favor, it suffers greater harm to its interests than any stilt injured while performing broken-wing behavior, or any bee that dies to bar the way of a hive invader. These latter organisms may, in death, satisfy many salient biopreferences. By contrast, goals that are among the most important for the warbler must remain forever unsatisfied.

It is important to keep in mind the differences between these other-requiring mate goals and the other-directed goals associated with kin-selected altruism. The relevant preference here is not targeted at the good of the mate. Any other-directed component of this biopreference will be targeted at the goods of potential offspring.

We can now find a different individualistic answer to Sober's $n + m$ question. Imagine we have a choice between two options. Option A involves the killing of five of the endangered takahe. Option B instead involves the deaths of five of the relatively plentiful pied stilt. Let's assume that both birds are very similar in terms of their levels of representational sophistication. This means that both options will share elements of harm. However, when we look beyond the individuals killed, we recognize that option A involves much greater damage to individual interests than option B. This is damage to the other-requiring goods of other members of the birds' populations or species. As pied stilts are plentiful, killing five will have no real impact on the ability of the remaining individuals to mate successfully. By contrast, killing five takahe will have a serious impact on the chances of other takahe to breed. Further, a small reduction in the ability of current takahe to produce offspring will translate into a much greater reduction in the chances that their offspring will produce offspring.

What if we are talking about the individuals belonging to a tiny remnant of a species? A great deal of public interest in New Zealand has focused on the black robin, once officially classified as the rarest bird in the world. In 1976 its total population consisted of seven individuals, including only two breeding pairs. Very skillful intervention boosted this to a 1992 figure of 128 individuals.[30] It seems that the approach I advocate will not assign much value to this very small population. As we have so few individuals, extinction will presumably involve the frustration of a correspondingly small set of selected goals. I think that recognizing the both other-directed and other-requiring nature of the goals of cognitively simple organisms allows us to see that allowing these last individuals to die may involve the generation of much more individual harm than killing two individuals belonging to plentiful species. As I have discussed, a common way of boosting inclusive fitness is by caring for relatives. Individuals whose psychology is shaped in this way will have biopreferences involving the care for genetic relatives. Two individuals belonging to a healthy population need not have these biopreferences decisively frustrated in death. The actions of others can lead to the satisfaction of many of their goals. By contrast, the last two individuals of a species will have their environment-directed goals utterly thwarted in being killed.

Our search for content-characterizable goals should not end with those of the last living black robins. I think we should also consider the goals of the expired relatives of the individuals belonging to this remnant. In chapter 6 I have shown that there was a difference between goals generated with respect to sentience, or sentience-goals, and selected goals. Sentientists say that a goal is important only insofar as it has conscious effects. Biopreferences have no such link with consciousness. They can be satisfied when the targeted effects are too spatially distant to be detected by their bearer. A parent's concern for the well-being of offspring survives even when they are separated by many kilometers. Distance also has a temporal dimension. Some biopreferences can be fulfilled by effects that occur after the death of their bearer. Though we should probably say that the preferences of individuals "fade out" after a few generations, there is no reason to see them expiring instantly on the point of death. In the case described in the last paragraph, the two remaining individuals can carry the hopes of great numbers of their expired conspecifics.

I have argued for a conception of individual interests that involves other organisms. Does this suffice for the assignment of intrinsic value to species? As we have already seen, there is a complex definitional issue here. *Species* is a contested word. According to my account, value will attach via the concern for the goals of individuals falling under a Mayr's biological species concept. An individual's interests extend to all other organisms that are relatives and, beyond, to all of those that it might mate with and those that might produce offspring that its offspring might mate with. So, the individual's interests mark the boundaries of the Mayrian biological species to which it belongs. My individualistic preservationist ethic, therefore, protects biological species.

The closeness with which my extension of individual value tracks biological species allows us to resolve species-concept conflicts. Take the problem discussed above of the threat posed to the black stilt by its tendency to mate with pied stilts. We saw that holistic species ethics failed to offer guidance. The individualist ethic does.

My analysis approaches the apparent threats of predation and cross-species interbreeding in very different ways. As the union of black and pied stilts produces offspring that are fertile, the interests of individual black stilts can extend to cover local pied stilts, and vice versa. Anthropocentric grounds will need to be introduced to motivate keeping the black and pied stilts apart.

I have argued that we can understand the concerns of species preservationists individualistically. The value of an individual within a species is not a function only of its own goals. Its demise affects the other-directed and other-requiring goals of its conspecifics, depending on how plentiful its species is. This makes individuals belonging to endangered species more valuable than those belonging to nonendangered species.

How to Value Ecosystems

I have shown how the concern for individuals can lead to a concern for species. Ecosystems are collections of individuals posing quite different problems to the ethicist. I have argued that ecosystem value holists are unable to help us in conflicts between the different markers of ecosystem flourishing. A focus on the other-directed and other-requiring goals of individual organisms does this job for us. We do best in individualistic terms by preserving the level of species richness in an ecosystem.

I have just shown how the interests of an individual extend to its conspecifics. For members of endangered species, each death of a conspecific

reduces the chances of reproducing. We can make distinctions among conspecifics, however.

Take the following case. The New Zealand tuatara is amongst the most
unusual reptiles in the world. It belongs to the ancient order of Rhyncho-
cephalia, which predates the dinosaurs; among its more bizarre adaptations is a partially functional third eye. The tuatara fared no better against
rats than many New Zealand birds and has become restricted to island
refuges.

Biologists used to group all tuataras in the same species. However, it is
now recognized that the tuatara has two species, *Sphenodon punctatus* and
the rarer *Sphenodon guntheri*. Until recently the last *S. guntheri* population
was to be found on a four-hectare island between the North and South
Islands. Any species restricted to such a small area is at the mercy of some
natural disaster or incursion of rats. To avert the annihilation of the species,
conservationists decided to incubate some *S. guntheri* eggs and establish a
second population on a large rat-free island.[31] Now, as very recent ancestors
of the large-island tuataras will be found on the smaller island, we should
certainly say that the populations on both islands are of the same species.
Small-island individuals would certainly interbreed with large-island individuals, given the chance. Despite this, as conservationists impose a reproductive barrier, they also create a barrier in individual interests.

We need to ask how the death of an individual on one of the islands
affects the goals of individuals first on the same island, then on the other.
Small-island individuals are very unlikely to mate with large-island
tuataras. So the death of a small-island individual does not affect the reproductive prospects of large-island *S. guntheri* in nearly the same way that it
does the reproductive prospects of its island-mates. An individual's interests extend most strongly to conspecifics belonging to the same local population.

We can now answer the ecosystemic version of the *n* + *m* question. The
deaths of five individuals belonging to a population whose tenure in an
ecosystem is secure will lead to the frustration of fewer individual goals
than the deaths of five members of a population on the verge of being driven out of an ecosystem. This is because it will have less impact on the
reproductive prospects of local conspecifics. So, we can motivate, in purely
individualistic terms, the assignment of greater value to organisms belonging to threatened local populations.

This reasoning allows us to answer questions about the importance of
properties such as diversity, stability, beauty, and integrity. If the equilib-

rium view about ecosystems is correct, we will have vindicated moral interest in these properties. Ecosystems that are stable, diverse, beautiful, and possessed of integrity will also have secure local populations. However, imagine that disturbance theorists turn out to be right. Frequent disturbance is all that maintains many populations. In this case, we may have good individualistic grounds for often disdaining stability.

Once we have an individualistically grounded concern for local populations, information about the structure of ecosystems will allow us to modify the value of organisms belonging to other species. Not all populations in an ecosystem will be equally important to the maintenance of overall species richness. Some could be lost with only minimal impact on other populations. The demise of others will have wider impact.

So my suggestion about how we could "in principle" calculate the value of an ecosystem is as follows: For illustrative purposes let's assume the simplest consequentialist function from the values of individuals to the value of the whole. This individual value is a function of individual representational complexity. We cannot just sum this value, as facts about the community and ecosystem to which the individual belongs require us to modify our initial assignments. Members of threatened populations will be given more value than they otherwise would. This is because they are essential to the flourishing of conspecifics within the ecosystem. If a population is especially important to ecosystemic health, then its members should also be viewed as more valuable. Individuals belonging to other species will depend disproportionately on them. The important thing is that all these qualifications of the value of organisms are motivated in individualistic terms.

In this chapter I have given an account of the values of species and ecosystems founded on the values of constituent individuals. In the final chapter I will show how the seemingly complex series of value modifications and additions demanded by life ethicists leads to a theory that human beings can follow.

CHAPTER 8

An Impossible Ethic?

BIOCENTRISM, CONSEQUENTIALISM, AND COGNITIVE TRACTABILITY

In this book I have aimed to show that challenges to long-held metaphysical views have implications for ethics. Folk morality founds its assignments of value on the folk metaphysical scheme for grouping objects into categories. Science redraws the boundaries of intrinsic value by offering a better story about morally relevant samenesses of and differences between objects. This better story comes in the form of theory about psychological natural kinds.

The directive to hunt out morally interesting natural kinds is not so straightforwardly complied with. With most value-endowing notions, we find that we are spoiled for choice as more than one kind emerges as a candidate for value. This is no surprise. Folk psychological notions have evolved without the benefit of detailed information about the boundaries of crisscrossing kinds. These multiple kind-overlaps may be an affront to enthusiasts of philosophical tidiness, but they represent an opportunity for those with revisionist ethical agendas. I have suggested that we count each candidate kind as a plausible naturalization of the folk psychological notion and assign value in proportion to the closeness of the kind to the original value-bearing notion.

Among the aftershocks of science's challenge to folk metaphysics is an environmental ethic. Though folk moralizers are disposed not to assign intrinsic value to stick insects and ocean pelagic ecosystems, once the boundaries of natural kinds are illuminated, we learn that such things do indeed qualify for some degree of this most important variety of value.

Thus far a summary of the main conclusions of this book. Can we really accept them? Owen Flanagan's "principle of minimal psychological realism" will serve as a useful starting point for a line of objection to attempts like mine to radically overhaul received ethical opinion. Flanagan offers his principle as a minimum requirement for a moral theory: "Make sure when constructing a moral theory or projecting a moral ideal that the character, decision processing, and behaviour prescribed are possible . . . for creatures like us."[1] Vagueness and inconsistencies notwithstanding, commonsense morality embodies requirements tailored to humans by millennia of folk experimentation. Proposed displacing ethics may be theoretically cleaner and supported by arguments that we are unable to refute. However, if we are incapable of accepting their guidance, they surely fail as moral theories, for the primary task of a moral theory is to guide behavior. Advice that is impossible for humans to follow is bad advice, regardless of how high-minded it might seem.

We must be careful not to go too far in meeting this objection. Biocentrists want a theory that is possible for humans to follow, but not so generous to us that we almost all turn out to be de facto biocentrists. In chapter 4 I have shown that oversolving the problem of demandingness is a risk for life ethicists. Pluralists make biocentric value less of a burden in a way that leaves open mere lip-service recognition of it. The value of living beings risks being entirely crowded out by the higher value of sentient or rational beings.

This chapter places biocentrism alongside consequentialism, the family of theories that is the most popular target of the demandingness objection. This will help me examine in greater detail two ways in which biocentric value might be thought to ill-fit human beings. First, there is the *cognitive* version of the objection. Here, the target theory is held to be inhumanly computationally complex when applied to practical situations. Daniel Dennett (1995) explains that a theory whose job is to guide behavior must be computationally tractable. To satisfy this requirement, there must be a feasible algorithm, or a series of steps, intelligible by humans, leading to the action prescribed by the theory. We could as well be guided by the conclusions of a theory that fails this test as we could use quantum physics to guide our interactions with coffee cups and train conductors.

Second, there is the *scale of burden* objection. Once we work out what is required by a theory that is the target of this variant of the demandingness complaint, we are supposed to find that so much is asked as to leave no room for any recognizably human life.

Consequentialism is supposed to be vulnerable to both these different versions of the demandingness objection. Biocentrism has similar apparent weaknesses.

Consequentialism and Computational Tractability

According to consequentialists, the consequences of an action alone determine its value; morality is concerned exclusively with outcomes. The best-known variant of the theory, utilitarianism, ranks outcomes in terms of their relative compositions of happiness and suffering. Suppose I have twenty-five dollars in my pocket. This sum of money presents me with a number of what Flanagan calls *action opportunities.*[2] I can buy a book that may help improve my bridge play; I can buy some flowers for a friend; I can deposit the money in a bank account; I can leave it in my pocket; or I can donate it to charity. The outcomes of these and countless other possible courses of action must be collectively scrutinized to see which maximizes happiness. Of the options described, giving to charity seems, at least initially, to be the one likely to bring about the biggest positive moral change.

So, how does the cognitive version of the demandingness objection represent a threat to utilitarianism? We can break utilitarian decision-making down into two stages. First, the utilitarian needs to determine what possible courses of action are open. Next, the utilitarian needs to evaluate each possible action opportunity, acting on the one that receives the morally best valuation. Opponents of utilitarianism have alleged that both stages are cognitively beyond human beings. Here is Flanagan on the first stage of the utilitarian decision-process:

> What is an action opportunity? Action opportunities cannot simply be opportunities one notices or takes, for there are surely many opportunities one fails to recognize or take advantage of. From a consequentialist point of view it will be best if one notices and takes all available good-producing action opportunities. How many such opportunities are there? There is no determinate answer to this question, but the indeterminacy occurs in some astronomically high range. Think of all the actions you might perform in the next five minutes, or even in the next thirty seconds. . . .[3]

According to Flanagan, this inability to account for options means we must abandon utilitarianism as a moral decision procedure.

Even were we up to this, we would be incapable of giving each individual option the right treatment. Here is Dennett:

> It is unlikely in the extreme that there could be a *feasible algorithm* for the sort of global cost-benefit analysis that utilitarianism (or any other

"consequentialist" theory) requires. Why? Because of what we might call the Three Mile Island Effect. Was the meltdown at the nuclear plant at Three Mile Island a good thing to have happened or a bad thing? If, in planning some course of action, you encountered as a sequel or probability p, what should you assign to it as a weight? Is it a negative outcome that you should strive to avoid, or a positive outcome to be carefully fostered? We can't yet say, and it is not clear that *any* particular long run would give us the answer.[4]

Considered in terms of its immediate effects, the meltdown is surely a bad thing. The utilitarian must look beyond these effects, however. Suppose the meltdown resulted in improved safety protocols at nuclear power plants around the world—and that these protocols prevented a nastier meltdown. Now Three Mile Island judged in terms of consequences is a good thing. More distant effects can easily reverse this verdict. The later meltdown might have allowed the formation of even better protocols that would have prevented an even more severe accident. And so on.

It follows that we cannot say whether the meltdown was a good, bad, or morally indifferent thing.

So, if the critics are to be believed, utilitarians can neither do adequate jobs of working out the range of alternative courses of action nor of appropriately evaluating any of them.

Biocentrism seems to have analogous weaknesses. Where utilitarians are alleged to face no end of options or of forecasting, biocentrists must take into consideration seemingly no end of variably valuable beings. Billions upon billions of living individuals, each weighted in accordance with its proximity to perfect deservers of value-endowing folk psychological concepts, need to be fed into our moral decision-making machinery.

To take the most mundane of examples: eating a hamburger improves prospects for some stomach bacteria while making much worse off many microbial meat-patty inhabitants. Were fast-food toxins to lower our resistance to infection, we might find a range of viral and bacterial beneficiaries. Another burger sold may, in some small way, further encourage the farming of cattle, allowing the identification of still more affected organisms. Once tracked down, every affected organism must be ranked in terms of its plausible folk psychologicalness. Further difficulties arise at this point. If we take a nonaggregative, nonmaximizing approach to biocentric value, we may content ourselves with having paid some heed to each of these trillions of separately biocentrically valuable things. On the other hand, an aggregative maximizing approach to biocentric value will require a truly fearsome series of value-additions and -subtractions to arrive at a decision about whether eating the burger is a biocentrically good thing.

The correct consideration of all the life-relevant implications of alternative courses of action would overwhelm a supercomputer, and it is surely beyond our abilities. Decision-making paralysis looms.

Biocentrists and utilitarians can grant many of the premises of the above arguments and reject their conclusions.

Utilitarians should concede that we are not faultless finders of morally optimal options. Humans are far from the beings an omnipotent and omniscient utilitarian God would have created. We can regret that we are not such beings without being excused of obligations to do what we can do. So although we cannot take account of every theoretically available action-opportunity, we may still be required not only to systematically choose the best options we happen to notice, but also to strive to become the kinds of beings who notice morally good options.

What of Dennett's Three Mile Island example? As described, it seems to pose insoluble problems to consequentialist evaluators. All the information available to even the most diligent utilitarian leaves open the possibilities that Three Mile Island was extremely good, that it was extremely bad, and that it was morally neutral. I propose that the apparent moral intractability of the 1979 meltdown no more invalidates consequentialism as a whole than do occasionally chaotically changeable meteorological conditions vitiate the entire enterprise of weather forecasting. Dennett will need to show that his diagnosis of the Three Mile Island problem applies across the board to estimates of future harms and goods. It strikes me that this would be the case only if we were to accept a near Humean skepticism about knowing the future. If so, we would need to take seriously the possibility that my purchase of a blueberry muffin today could lead to global nuclear conflagration tomorrow. Such extreme skepticism aside, we should be confident in making accurate enough judgments about the future good and bad effects of most of our action opportunities.

Some concession of epistemic and moral humility will also help biocentrists. We should concede that no human being is perfectly attuned to the vicissitudes of every living being. The moral agents created by an omnipotent and omniscient biocentrist God would be very different from us. This doesn't mean that the biocentric argument does not apply to us, nor that we cannot make good use of biocentric advice. Here's how we can begin to tackle biocentric complexities.

First, to the problem of the multiple degrees of value. In chapter 1 I have exposed the conceptual messiness of our familiar concept of intrinsic value. Different philosophical analyses pick out conflicting aspects of the folk concept, and there seems no way of deciding among them. These competing cashings-out of the traditional concept do tend to share one feature. They agree that intrinsic value is all or nothing; it does not come in degrees. The thesis of the moral equivalence of intrinsically valuable things may seem to make the task of deciding how to treat a thing relatively straightforward. Once we have decided that something is intrinsically valuable, there is no further issue as to how much value it has. It either qualifies fully for moral consideration or not at all. There is a downside to this lack of conceptual fuss, however. The thesis of the moral equivalence of intrinsically valuable things leads to some seemingly insoluble dilemmas. Upon learning that the weta and the black stilt have intrinsic value, we have no idea how to approach situations in which the interests of both conflict. Ad hoc maneuvers may help with specific cases, but only at some cost to the status of the concept of intrinsic value in our moral deliberations.

I have claimed in chapter 5 that fine discriminations of intrinsic value offer practical benefits. We can use degrees of value to guide us when we need to promote the interests of one intrinsically valuable thing at the expense of the interests of another. To reap such practical benefits, I must show that we are up to the task of tracking these many degrees. The key point here is that proximity to folk psychological paradigms is what determines how much value an object possesses, and we have cognitive machinery dedicated to the task of detecting and describing such states. Intrinsic value may be tied to complex natural features, but as our theory of this value is parasitic on our intuitive theory of mind, this complexity is tractable by us.

There certainly are conceptual and computational tangles that accompany our use of folk psychology. Prolonged philosophical dispute shows that it may often be difficult to work out whether or not a given organism is *genuinely* folk psychological. To carry this task off, we need to decide which parts of the overall theoretical package underlying our ability to predict behavior pick out states that are essential to folk psychology. We then make a judgment as to whether these essential parts are true of the organism under consideration.

Life ethicists ask that we undertake a much less conceptually tricky decision-making task. An acquaintance with a variety of organisms tells us, over time, what types of predictive strategies tend to work and which tend to fail. As we are not trying to use this information in a pass/fail way, we do not need to sort these strategies into the essential and the accidental. Rather, we aim to get a sense for how well, overall, we can predict and explain the organism and how much of folk psychology we need to do this. Our best guess of the percentage of the folk psychological package true of an organism will guide judgments about the proximity of the organism's behavior-guiding states to value-endowing paradigms.

There is good evidence for the psychological dispositions required by biocentrism. Consider our dispositions to empathize. We have no tendency to put ourselves in the place of a rock—or to recognize it as any kind of striving being. Though psychologists may insist that ants have no minds, we can empathize with them to some degree. Certainly it would be a mistake to superimpose on the ant the subjective experiences I might entertain while trying to roll uphill an object whose weight greatly exceeds my own. However, it is not mistaken to recognize the ant as seeking to do something with an object that stands in the same relation to it and, given the story told in chapter 6, not wrong to do so. Empathy gets easier as organisms get more complex. So value increases.

Now to the problem of number. Can we render computationally tractable the multitude of the alive? Again, I think the answer is yes. Chapter 7 has addressed the supposed metaphysical conflict between individualistic and holistic value. I have argued that we did not need an independent value-holism to morally motivate species and ecosystems. The appropriate concern for species and ecosystems tends to do best by their constituent individuals. An answer to the supposed epistemic inaccessibility of individual value can be found in the solution to the metaphysical problem. Individual value is metaphysically prior—the approach I have opted for dispenses with any holistic value that cannot be accounted for in individualistic terms—but holistic value is epistemically prior. Though our fundamental moral accounting is in terms of individuals, a focus on species and ecosystems enables us to *summarize* individual value. Such collectives perform this task much better than any other similar-sized collections of individuals.

DOES LIFE VALUE LEAVE ROOM FOR HUMAN LIVES?

A second concern arises over the scale of the burden biocentrism imposes on us. Good arguments for biocentric value would count for nothing if we could not reconstruct human lives around such value. Supposing we can work out how best to promote life-interests, would committing ourselves to such an ethic leave room for anything we would call a human life?

Again, let's first see how this type of objection runs against utilitarianism. Bernard Williams argues that utilitarians must lead lives devoid of true friendship and of nonmoral projects. They must exchange any such projects or personal attachments for the all-encompassing project of maximizing good, impersonally conceived.[5] Although sometimes acting in ways that seem consistent with these commitments, a utilitarian should be prepared to abandon them the instant utility calculations so dictate.

Biocentrism faces similar difficulties. Biocentric value seems so omnipresent as to make impossible the pursuit of any project not directly validated by the life ethic.[6] Innocent-seeming activities such as learning to play the violin and collecting movie memorabilia both impose costs on living creatures and fail to promote their interests. One's friends may be living and, therefore, worthy targets of our moral efforts, however much of what we do with them, activities that we take to be constitutive of friendship, seems dismissive of biocentric value.

Early utilitarians were inclined to bite the bullet when faced with this type of objection. They freely acknowledged that theirs was a revolutionary morality—likely to fly frequently in the face of intuitions about the good and just life. Biocentrists may also respond in this vein. Current intuitions about what would count as a worthwhile human life have formed in the shadow of human-centered moralities. This is bound to make the biocentrically informed life unappealing to many. We must remember that at various times, going to war, owning slaves, and acquiring estates in supposedly uncivilized lands were considered essential to the good life. Faced with the universalistic humanism popular today, the reactions of a thirteenth-century European nobleman, an ancient Roman patrician, or a Castilian conquistador may be similar to those of our contemporaries when confronted by the biocentric ethic. Earlier moral revolutions were not derailed by the complaint that a universalistic morality left no room for a life worth living, and so, too, biocentrists should not shy away from unpopular demands. Taylor doesn't pull his punches: "There should be no illusions about how hard it will be for many people to change their values, their beliefs, their whole way of living if they are sincerely to adopt the attitude of respect for nature and act accordingly. Psychologically, this may require a more profound reorientation."[7]

I don't want to rest too much on perhaps overconfident claims about the ultimate malleability of intuition in the face of revolutionary moral demands. Instead, I want to show that although the life ethic requires effort, it is not an all-consuming morality. Biocentric value need not drive out all competitors for our affections, and biocentrically good lives are likely to be recognizable as worthwhile lives.

Those utilitarians who see the need for some softening of their position typically urge that we distinguish between utilitarianism as a criterion of rightness and utilitarianism as a decision-procedure.[8] Sometimes we can achieve what the utilitarian would count as the best result by reasoning in ways that pay no heed to the principle of utility. For example, we may decide that the world is a better place with friendships in it and at the same time accept the criticisms of Williams and others that utilitarian deliberations are incompatible with friendship. We will then undertake not to be consciously guided by the principle of utility in thinking about our friends. To the extent that it is present, the principle will be well in the background of our practical reasoning. I will spend no more time on this keenly contested debate, as I claim that biocentrists have a more straightforward and direct line of response to the scale-of-burden objection. Even if we keep biocentric value in the motivational foreground, we can still live recognizably human lives. Instead, I will argue for a seemingly paradoxical result. It may be that biocentrists argue for a dramatic proliferation of intrinsic value, but the resulting ethic is *less* burdensome than some familiar ethical positions with narrower value bases.

It is initially tempting to think that there should be a straightforward correlation between the demandingness of an ethic and the number of things to which it assigns value. Ethics with narrow value bases require us to look out for relatively few things. As we broaden the value base, more objects must preoccupy us, and so the ethic becomes more burdensome. This correlation seems, at least initially, to hold up as we move from human-centered commonsense morality to Singer's animal-welfarist ethic, for example. We can meet almost all the requirements of conventional morality by treating other human beings appropriately. Singer holds that beyond what we owe to other human beings is a whole range of animal-directed duties. So, even if we were to forget for a moment that he has a much more demanding universalist understanding of what counts as correct treatment of other humans than does commonsense morality, it seems clear that Singer asks more of us than does unadulterated moral common sense.[9]

Appearances notwithstanding, there turns out to be no simple correlation between the size of the category of intrinsically valuable things and the demandingness of an ethic founded on this category. In the next pages

I compare the burden imposed by animal welfarism, as exemplified by Singer, with that imposed by the biocentric ethic outlined in this book. Although the set of things held by biocentrists to be intrinsically valuable is much larger than the set of things similarly valued by animal welfarists, animal welfarism turns out to be more demanding than the life ethic.

There are factors beyond the number of would-be intrinsically valuable things that bear on how burdensome an ethic is.

As we have already seen in this chapter, one issue with implications for the demandingness of an ethic is the broader ethical framework in which we choose to lodge our intrinsic value. If we choose a consequentialist framework, then we should expect our ethic to be rather more demanding than it would be were we to choose other frameworks. The consequentialist biocentrist requires not only that we refrain from harming living things without a morally defensible reason, but also that we actively promote life interests. So, beyond taking care not to step on ants, we must strive to make the world a place where ants are less likely to be tread on. Other ethical frameworks may require an attitude of respect for living things. This respect for life may rule out careless treadings-on but will not always demand the active promotion of life-interests. As I have argued in my chapter 4 discussion of Taylor, it seems to be much easier to respect each individual living thing than to go about actively promoting its ends. It is beyond the scope of this book to resolve the debate between consequentialists and their critics. For argument's sake, I will lodge biocentric value in the most demanding maximizing consequentialist framework.

Fragility and Robustness of Ends

To swiftly recap: Singer requires that we give equal consideration to the interests of all sentient beings. In my chapter 2 presentation of the view, I have suggested that "sentience" picks out a natural kind encompassing all organisms capable of experiencing suffering or pleasure. The natural kind provides justification for and defines the limits of this equal consideration. In our dealings with sentient beings, both human and nonhuman, we must be careful to cause as little suffering and as much happiness as we can. The question will be, how does this theory compare with a consequentialist biocentrism requiring that we act so as to bring about the best possible balance of biopreferences satisfied over biopreferences thwarted?

It is clear that the biocentric moral universe is larger than that of the animal welfarist. There are two reasons for doubting the move from the claim that biocentrists find more things intrinsically valuable to the idea that biocentrism is more burdensome.

In making judgments about demandingness, much depends on how difficult our selected ethical theory makes the promotion of or respect for its favored objects' ends. For some theories, the ends of objects falling into this favored category are *fragile*, while for other theories, ends are *robust*. If the ends of our favored objects are fragile, then an ethic centered on them will be somewhat demanding. If they are robust, the ethic will be comparatively undemanding.

In some cases, this difference can be traced to the selection of objects, with fragile-end ethics placing importance on structurally fragile objects, and robust-end ethics favoring more structurally robust objects. Compare an ethic assigning intrinsic value to sand grains with one assigning such value to ice crystals. Suppose the interests of both sand grains and ice crystals are promoted or respected by ensuring or allowing that they maintain their structures. In a world similar to our own in terms of physical laws, the promotion of or respect for individual ice-crystal ends should demand more of our energies than would the promotion of or respect for individual sand-grain ends. Such considerations begin to disrupt any simple correlation between numbers of objects valued and how burdensome an ethic is. Imagine that there are exactly one hundred ice crystals and one thousand sand grains, all scattered about indiscriminately. We would expect the ice-crystal ethic to be more demanding than the sand-grain ethic, in spite of the disparity in numbers of intrinsically valuable things involved.

The tendency of different theories to focus on different properties of the same objects will also make for different judgments about the fragility of ends.

Think about the ease with which an animal can be made to suffer. Though animal welfarism's principal advertised concern is the suffering generated by such human activities as factory farming and vivisection, since the biological purpose of pain is to equip an organism to deal with recurrent natural hazards, there are no surprises that it should be omnipresent in nature. Value as identified by the animal welfarist threatens to become truly overwhelming if, as indeed Singer does, we choose to lodge it in a consequentialist framework. It seems hard for Singer to ignore the calls of sentient herbivores threatened either by nonsentient predators or by sentient predators with nonsentient prey-alternatives.[10]

The life ethic fares better in this respect. An interest in organisms guided by a concern for the satisfaction of their biopreferences makes their ends comparatively robust. Sentient beings can suffer enormously and yet satisfy biopreferences related to important activities such as reproduction and the protection of offspring. Indeed, this is typical. A Thomson's gazelle eaten to provide fuel for a cheetah's reproduction is not debarred from having satisfied significant biopreferences. In terms of biopreference satisfaction, there may often be very little to choose between gazelles eaten and gazelles not eaten. Cheetahs often catch weaker animals, those that are beyond reproductive years, or so infirm as to have slim reproductive opportunities anyway.[11]

In this respect, at least, Singer imposes a greater burden than does the biocentrist. Returning to the issue of number, we can find further grounds for thinking the biocentric burden more bearable than the animal-welfarist one.

From Value Eliminativism to Value Everythingism

There is a point beyond which further increasing the size of the set of intrinsically valuable things tends to produce a less demanding ethic. The reduction in burden occurs as the world becomes crowded with intrinsically valuable things, in turn generating a phenomenon that I call *end-interdependence*. This phenomenon will tend to make biocentrism less demanding than animal welfarism.

To illustrate this relationship between number, end-interdependence, and demandingness, I will trace a path from ethics assigning intrinsic value to few things, and hence distributing value sparsely around the world, to those assigning intrinsic value to many things, and hence making the world crowded with value.

At one end of the spectrum of possible moral positions is a variety of moral eliminativism. Those who hold this position deny that there is anything that is intrinsically valuable. Recent writers on the environment who find debate on the meaning of "intrinsic value" bothersome may count themselves as intrinsic value eliminativists.[12] However, the position I am imagining is rather more extreme than any to which they would subscribe. I have taken the concept of intrinsic value to be relatively malleable, free to track wider theoretical commitments. According to this view, not only will there be nothing that is intrinsically valuable, but there will be nothing falling under any concept filling the same broad moral-theoretic role as "intrinsic value."

Unless we can find some other ground for moral duties and responsibilities, this ethic, if indeed it deserves such a name, turns out to be mini-

mally demanding. We are denying that there ever are any morally significant interests or needs to be taken into account.

As we move away from the value-eliminativist extreme, it makes sense
that we find progressively more demanding ethics. The positions we will
be considering direct us to look out for an increasing number of things,
and all else being equal, it is plausible that more effort must be put in.
Think of the difference between the parochial humanism of commonsense
morality and the universalist humanism recently defended by Peter Unger
(1996). While miles more demanding than intrinsic value eliminativism,
the former view requires only that we take an interest in humans near to
us. Beyond our circle of friends, relatives, and colleagues, demands fall off
precipitously. We owe almost nothing to the distant starving. By counting
in the distant starving, the universalist view becomes more burdensome.
Unger describes the changes one must make to one's life to meet the needs
of the world's starving. We must take up professions enabling us to generate as much money as possible to help alleviate global poverty and, indeed,
can hardly justify a ten-minute break to pursue any interests of our own.

Continuing our progression, we reach a point beyond which further
boosting numbers has no effect, because our motivational capacities are
already exhausted. If we are already overtaxed by the number of sea shells
whose ends we must respect or promote, adding five hundred thousand
morally valuable pebbles will make little difference to us.

While the move from universalist humanism to animal welfarism represents a straightforward addition of intrinsically valuable things, it is possible that animal welfarism is no more burdensome than universalist
humanism. If Unger's recent interpretation of universalist humanism is
correct, we may have already exhausted our motivational energies.[13]

There is a region of possible ethical space of greater interest to me. This
region is occupied by theories that assign interdependent ends to objects.
Two objects have interdependent ends if the achievement of the first
object's end bears on or is borne on by the achievement of the second
object's end. Systematic end-interdependence could occur when the world
is so crowded with intrinsically valuable things that many of them are
physically adjacent. Two varieties of interdependence are of interest. The
first occurs when ends of objects are *reinforcing*. Here, in seeking to achieve
its end, object *A* will tend to contribute to object *B*'s end, and possibly vice
versa. In chapters 4 and 6 I have discussed one way in which the ends of
organisms as understood by the life ethic are reinforcing. They can be
other-directed—having ends targeted at the ends of other organisms. The

other side of end-interdependence is *conflict*. Here, the ends of objects tend to exclude one another; the achievement of *A* ends rules out the achievement of *B* ends, and vice versa.

I suggest that both types of end-interdependence, if widely enough found, tend to make an ethic less demanding. Systematic end-reinforcement does this by making objects' ends more robust than they would otherwise have been. Nearby things will be doing the job of promoting or protecting a given object's ends, and human would-be interveners will need to decide whether we could do better by those ends than the reinforcing object. We must bear in mind that intervening may harm an object in its supporting role. In the last chapter I have described the conflict between the disturbance and equilibrium views of ecosystems. These views frequently dictate different kinds of intervention to those interested in aiding an ailing ecosystem. Budiansky (1995) describes many cases in which conservation efforts, when informed by mistaken ecological theory, can have effects opposite to those intended.

Widespread end-conflict has a similar effect. Suppose that *A*'s ends and *B*'s ends matter roughly equally to us and that *A*'s and *B*'s ends cannot be achieved together; it may be that the frustration of *A*'s ends tends to be instrumental to the fulfillment of *B*'s ends. Think of predator-prey relationships. The knowledge that *A*'s ends are being frustrated by *B*'s will not demand action from us unless we can come up with some way in which both *A*'s and *B*'s ends can be fulfilled.

The tendency for crowded ethics to generate end-interdependence and thereby be less demanding becomes apparent if we examine an ethical position lying at the opposite extreme from the earlier discussed intrinsic value eliminativism. This theory assigns intrinsic value to everything; all actual and potential arrangements of matter are held to be intrinsically valuable. In the purest version of the theory, each arrangement of matter has equal intrinsic value.

Before assessing how demanding this value-everythingism is, I note that such a theory may be more than an idle possibility. The early sections of Roderick Nash's history of the intellectual foundations of environmentalism describe "sequential" ethics. Adherents of sequential ethics arrange current and past moral views into a linear evolutionary history represented by a cone-shaped diagram. Movement from the narrow to the wide end of the cone correlates with increasing numbers of things recognized as morally considerable. In Nash's "pre-ethical past," only the self is recognized as valuable. The section designated "ethical past" progresses from family through tribe to region as objects of moral concern. The contemporary era has witnessed the move from parochial or racist ethics to a recognition that all humans are valuable. Most recently, animal welfarists seem on the verge of bringing about a further enlargement.

The swelling of the moral community over the past centuries prompts this line of inductive reasoning. If the moral future is anything like the moral past, then surely next must come species and ecosystems. The most far-sighted of the inductive arguers wonder whether there might come a time when the entire universe and perhaps everything in it is recognized as intrinsically valuable.

There are a couple of questions we should ask about sequential ethics. First, how does this story relate to any real facts about human moral history? Certainly, as Nash concedes, it is at best an idealization. For a start, there is unlikely to have been a stage at any point in human moral history when only the self was recognized as valuable. Frans de Waal (1996) points out that we share with apes and monkeys the kin-selected and reciprocal

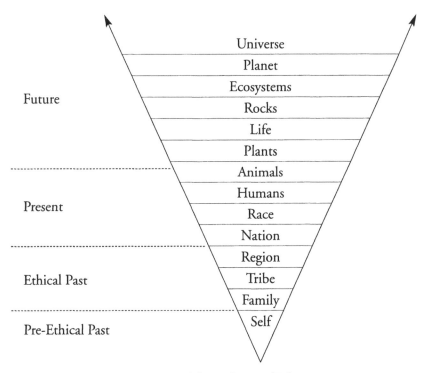

FIGURE 8.1 *The Evolution of Ethics*

altruisms forming the foundations of our moral instincts. This surely indicates that they have been present since well before there were humans. Furthermore, it is unlikely that there was anything like the smooth progression that Nash's diagram indicates. Even today, universalist humanist ethics jostle with nationalist, racist, and other more parochial moral relics.

Idealized histories are certainly not unheard of in contemporary moral and political philosophy. Most famously, John Rawls (1972) has described an original position from which principles of morality are chosen. Contractors in this original position are supposed not to have any information about their genders, natural endowments, or positions in society. This eliminates the class, race, status, or other biases that drive many of our actual choices, enabling the contractors to arrive at principles that are fair to all. It doesn't matter that no historical human society even remotely resembled this situation. We need only agree that it constitutes an appropriate starting point for the formulation of just principles to govern a society.

Rawls's and Nash's theories are radically different in aim. The most glaring difference is that Rawls's is not an inductive argument. Idealizing does create problems for inductive arguers. If the future is supposed to be like the actual rather than the idealized past, then the more differences we find between the actual human past and Nash's story, the less faith we should place in his depiction of the future.

Put the status of Nash's argument to one side. Suppose we do arrive at an ethic that recognizes intrinsic value in every pattern of matter. Despite its huge value base, our value-everythingism is very undemanding, indeed. We need only think of the profusion of end-conflicts. Though every readjustment of matter destroys some intrinsically valuable pattern, as it does so it must bring into existence an equally valuable arrangement, or protect an arrangement that would have been disrupted. This end-interdependence means that we never have any reason either to intervene, or to curtail an activity, on behalf of intrinsically valuable things. As in the case of intrinsic value eliminativism, morality will not constrain behavior at all.

Thus far, this is a somewhat abstract and schematic discussion of the relationship between the size of the set of intrinsically valuable things and the demandingness of an ethic. We can now return to our comparison of animal welfarism with biocentrism.

While large, Singer's value base is not so large that interests are systematically interdependent. His is not a particularly crowded ethic.

There is, of course, for the animal welfarist, some interdependence of ends. The lioness tends to suffer when her cub does, and this gives her a

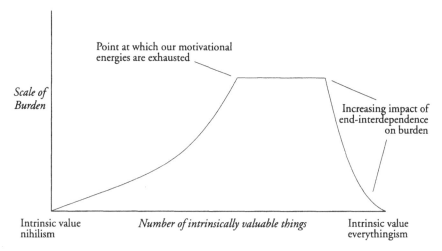

FIGURE 8.2 *The Relationship Between Numbers of Intrinsically Valuable Things and Scale of Burden*

reason to look out for the welfare of that cub. Although this is a clear example of end-reinforcement, such cases will be limited to networks of kin-selected sentience-preferences. Few occupants of most ecosystems are sentient, and the animal welfarist must ignore the vast networks of unconscious end-reinforcement.

There are also end-conflicts to which the animal welfarist will appeal, and these will occasionally reduce the moral burden. Details of the manner of death standardly inflicted by lions on zebras would horrify animal welfarists. As if the terror of the chase where not enough, the jaws of the lion are not sufficiently powerful to crush the neck of the zebra, killing it instantly. Instead, the recent capture must endure five or more minutes of asphyxiation as the jaws clamp around its windpipe. All this while other members of the pride rip still-living flesh from the zebra's haunches and tear open its gut to feed on its viscera. This all seems very bad, yet balanced on the other side of this sentientist's nightmare we understand that the sentient lion must eat the sentient zebra in order to avoid a longer-played out, though less intensely painful, death by starvation. Until we have a lentil-gazelle capable of fulfilling both the lion's need for nourishment and hunting, animal welfarists need not beseech Serengeti nature-documentary makers to intervene in the hunt.

Other cases of apparent end-conflict, however, seem straightforwardly to call for intervention by the animal welfarist. Singer's moral accounting

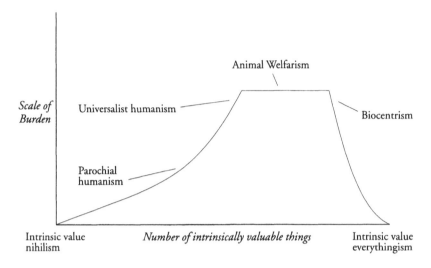

FIGURE 8.3 *The Burden of Biocentrism Compared with Other Moral Positions*

makes it fairly certain that rabbits should not be tormented in the testing
of cosmetics. Human preferences satisfied by the search for novel scents
are either satisfiable in ways that cost less in terms of animal suffering or, if
not, are minor in comparison with that suffering. By the same token, it is
at least conceptually clear that army ants ought not capture and consume
rodents nor spiders' webs snare small birds. Any spider or ant ends are supposed to merit no independent representation in our moral reckonings.
Singer says that we cannot blame the ants or the spiders, for they are not
moral agents.[14] Be this as it may, worlds in which webs fail to snare birds
and rodents escape are worlds in which, other things being equal, more
preferences of sentient beings are satisfied, even if army ants go hungry
and spiders are restricted to nonsentient prey.

  Once we take into account Singer's preferred consequentialist moral
framework, the fragility of animal ends, and the fact that though the set of
intrinsically valuable things is large, its impact on us is not mitigated by
systematic end-interdependence, we recognize how distant and saintly is
the life of the truly conscientious Singerian animal welfarist.

  The life ethic is sufficiently expansive that the interests of the intrinsically valuable are systematically interdependent. The natural world is a
vast network of sometimes coinciding, often conflicting biopreferences.
Focus on kin selection and reciprocal altruism provides us with a range of
reinforcing ends. In contrast to Singer, however, these extend far beyond

the category of the sentient. Widespread end-conflict further helps to reduce the burden. For each biopreference frustrated, we should expect to find biopreferences satisfied. When heron nestlings fall into the waiting jaws of Floridian alligators, there is a cost according to both the animal welfarist and the biocentrist. The heron experiences pain and is also unable to satisfy many biopreferences. Much more frequently than animal welfarists, life ethicists will find things to balance against these frustrated biopreferences. In this case, we can note that were it not for patrolling alligators, great numbers of heron eggs would become food for raccoons. Though they are moral zeros for the animal welfarist, the biopreferences of eggs are every bit as visible to the biocentrist as are those of the chick.

Considerations about end-interdependence and robustness of ends provide good grounds for thinking that biocentrism is less demanding than the widely advocated animal welfarism of Singer. What manner of achievement is this? After all, Singer's is a moral view famed for its demandingness. I certainly do not want to deny that the life ethic is burdensome. In these last paragraphs I propose that, however large the biocentric burden, we should accept it. This is because it is largely of our own making.

Though for biocentrists the world is fairly crowded with intrinsically valuable things, the degree of end-interdependence is clearly significantly less than it is for our maximally undemanding moral everythingist. Even in places saturated with life, there are things we can do that frustrate biopreferences without realistically boosting the chances that other biopreferences will be satisfied. Highly toxic factory effluent may eliminate all living things in an area, at the same time rendering it uncolonizable by almost any other organism. I suspect that there was a time when the practical differences between moral everythingism and biocentrism would have been small. In chapter 2 I have discussed Wilson's evolutionary argument for an environmental ethic. I claimed that our attitudes to nature, if in fact they have been significantly shaped by natural selection, are likely to have been designed for a Pleistocene world in which a thoroughly exploitative attitude to nature had no bad consequences for humans. I want now to use the language of the life ethic to make a related point. Even had they been capable of grasping it, biocentric value would have made very few demands on these one hundred thousand hunter-gatherers. This is because the humans of that era would have found it difficult to act in ways that significantly ruptured the network of interdependent biopreferences. For any victim of human activity, there are likely to have been many natural beneficiaries. Though such calculations could only ever be made very approximately, biopreferences thwarted are likely to have been roughly balanced by biopreferences promoted.

The modern human skill in coming up with activities that impose great costs on nature is responsible for the moral burden we currently face. Were we not to have done so much to damage and destroy species and ecosystems, we would need to do comparatively little to treat nature right. So, although today we find that the burden imposed by biocentric value is fairly substantial, for this we have only ourselves to blame.

The first seven chapters of this book are a sketch of a biocentric ethic—an ethic that finds some degree of intrinsic value in every living thing—and an argument that this ethic affords adequate moral protection to nature. This chapter has had a more practical focus. My question has been: Can the life ethic guide the behavior of individuals who are not only moral agents but also human? My approach to practical matters has been somewhat indirect and theoretical. I have not given a list of applied biocentric principles and pointed to agents who exemplify them. Nor have I described biocentrically correct conservation programs and argued that the resources and political will exist to implement them. Instead, I have argued that applied principles, when they are described, are unlikely to be cognitively and motivationally beyond us.

Some will be impatient with this apparent caution. They will say that environmental ethics starts with a recognition that there is an environmental crisis caused, in large part, by humans' blindness to nature's moral worth. I may have demonstrated the genuineness of environmental intrinsic value but, beyond condemning the most obviously destructive human activities, I make little effort to describe the behavior this value entails. In not telling ordinary folk precisely what to do, I am quite simply abdicating my responsibilities as an environmental ethicist.

My caution stems, in large part, from my philosophical naturalism. According to this approach, philosophy does not stand imperiously apart from science, making rules with which scientists must comply. It is therefore not for me to lay down the moral law in advance of the relevant empirical work. In lieu of precise instructions, I offer a reconceptualization of the sciences of nature that will have immediate and far-reaching practical implications. The sciences of nature and the ethics of nature are no longer separate disciplines. Finding out about natural kinds constituting the environment enables us to see not only how the environment might be protected, but also that it should be protected. Pointing to this conceptual tie between the scientific and the ethical appreciation of nature can recruit thousands of environmental scientists to the moral cause of the environment.

NOTES

⁂

1. THE PSYCHOLOGICAL VIEW OF INTRINSIC VALUE

1. See Gibson, McKay, Thomas-Keprta, and Romanek (1997).

2. See Jakosky (1996).

3. See Hoyle and Wickramasinghe (1993).

4. The above paragraphs have been purposefully sketchy. However, I may have already said enough to fall into serious error. Later in this chapter I will tackle these definitional issues. For now, I let the identifications of anthropocentric ethics with instrumental value and nonanthropocentric ethics with intrinsic value stand.

5. The concept of intrinsic value I have described is tied to the concept of life, and this will make it blind to some environmental wrongs. For example, half a century of climbing expeditions have littered the upper reaches of Mount Everest with detritus ranging from oxygen canisters to human bodies. As this despoliation has a negligible impact on living organisms, we will need to turn to the anthropocentrism of a theorist like Norton to show why it is really wrong.

6. See Schrader-Frechette (1995) for a discussion of this disciplinary split.

7. Rolston (1988), Callicott (1989), and Naess and Rothenberg (1989) are proponents of environmental intrinsic value. Hargrove (1992), Norton (1992), and Regan (1992) are recent critics.

8. See Fodor (1983).

9. O'Neill (1992) is an excellent survey of different approaches to intrinsic value.

10. Moore (1922:260).

11. Moore's and very likely Korsgaard's notion of intrinsic value will certainly raise the eyebrows of some environmental ethicists. It is a standard claim that rarity boosts the intrinsic value of an organism. A single giant cave weta is much more valuable than a single cockroach regardless of any similarities or differences in nonrelational properties. The fully relational property of rarity explains the difference. If so, environmental intrinsic value cannot be a Mooreian intrinsic value.

12. See, for example, Naess and Rothenberg (1989).

13. See Elliot (1997:chapter 1) for a detailed exposition of this type of view.

14. See Sheldrake (1981).

15. See Lewis (1994) for this type of story about conceptual analysis.

16. Goodpaster (1978) is alert to the prospect that conceptual analysis of ethical notions may track entrenched social prejudices.

17. For discussions of theoretical definitions and their connections with natural kinds, see Millikan (1989b) and Papineau (1993:chapter 3).

18. Tooley (1972) is well known for this type of approach.

19. Quoted in Nash (1990:115).

20. Nagel (1974) and Jackson (1991).

21. Mill (1861:210).

2. Science's Bridge from Nature to Value

1. Taylor makes a similar suggestion. His arguments will be addressed in chapter 4.

2. Lewontin (1992:8).

3. Lewontin (1992:9).

4. Quoted in Allaby (1995:24).

5. Appleyard (1992:5).

6. Appleyard (1992:5).

7. See Erlich and Erlich (1994). Allaby (1995) is an effective response to Appleyard.

8. Appleyard (1992:124).

9. Appleyard (1992:137).

10. See, for example, Boyd (1988), Brink (1989), and Railton (1986).

11. See Sterelny (1990) and Dupre (1993).

12. For the best-known expositions of this view, see Churchland (1979), Stich (1983), and Churchland (1985).

13. For example, Jackson and Pettit (1989), Fodor (1987).

14. See DeGrazia (1996) and Varner (1998) for recent vindications of animal minds.

15. M. Dawkins (1990:4).

16. See also Rachels (1990) and Singer (1993).

17. See some of the essays in Singer and Cavalieri (1993).

18. Grumbine (1992:21).

19. White (1967:1205).

20. White (1967:1205).

21. Wilson (1994b:84).

22. Singer (1993:277).

23. Callicott (1994:152).

24. For a recent discussion, see Darwall, Gibbard, and Railton (1992).

25. Note that we will need a fairly hard-line version of naturalism. Ethical properties will need not only to be natural properties, but natural properties accessible to science. More moderate ethical naturalists might draw a parallel with one widely held view

about the relationship between the psychological and the natural. According to this view, psychological facts are natural facts, but they are not accessible to science in the way that facts of chemistry and physics are.

26. Safina (1995).

27. Wilson (1992:348).

28. Wilson and Kellert (1993:31–32).

29. Wilson (1992:349).

30. Wilson (1992:350).

31. Wilson (1992:350).

32. For some examples, see the papers in Barkow, Cosmides, and Tooby (1992).

33. Inclusive fitness is a wider notion of fitness calculated from an individual's own reproductive success in addition to his/her effects on the reproductive success of relatives weighted by the appropriate coefficient of relatedness.

34. See Sterelny (1995) for a discussion of the obstacles in the way of theories about psychological adaptations.

35. See the collection of papers in Thompson (1995).

36. Wilson (1979), especially chapter 9.

37. Ruse (1995:242).

38. Ruse (1995:257).

39. Ruse (1995:257).

40. Ruse (1995:283).

41. Ruse (1995:254).

42. Ruse (1995:281).

43. Wilson (1992:350).

3. Overlapping Kinds and Value

1. For a discussion particularly relevant to dissociative disorders, see Hacking (1995:chapter 7).

2. Devitt and Sterelny (1987:73).

3. Neander (1991b:174). See also Millikan (1989b) and Wright (1973).

4. Bigelow and Pargetter (1990:335).

5. Davidson (1987:443).

6. See Millikan (1989b) for a discussion of Swampman.

7. See Godfrey-Smith (1994) for a helpful overview of the functions debate.

8. See Medawar (1957) and Nesse and Williams (1995).

9. Midgely (1979:439).

10. Midgely (1979:446).

11. Turing (1950).

12. See, for example, Block (1981).

13. See Jackson and Pettit (1989) and Lewis (1994) for functionalist accounts of intelligence.

14. For a relevant discussion, see Fodor (1986) where he asks what kind of sophistication an organism requires for its behavior to be the product of its own representations.

15. For an argument against animal belief, see Frey (1980).

16. Stich (1992:243).

17. This approach is clearly outlined in Cummins (1989).

18. Schull (1990:63).

19. For a discussion of functionalism and qualia, see Lycan (1987).

20. Editorial commentary on Schull (1990:94).

4. Recent Defenses of Biocentrism

1. Schweitzer (1953:188).

2. See Nash (1990:60–62) for an economical and helpful description of Schweitzer's life ethic.

3. See, for example, Kekes (1993:121–22).

4. For a helpful, nontechnical introduction to Aristotle's views on life, see Ackrill (1981:chapter 5).

5. For a computational approach to the principles that make up commonsense physics, see Holland, Holyoak, Nisbett, and Thagard (1989).

6. See Holland, Holyoak, Nisbett, and Thagard (1989:chapter 7).

7. Taylor (1986) differentiates inherent worth from intrinsic value. He gives intrinsic value an anthropocentric reading. Roughly, an object is intrinsically valuable if it is valued by a human valuer for itself. An object possesses the properly nonanthropocentric-inherent worth by virtue of having a good of its own. Taylor's "inherent" should be seen as interchangeable with my "intrinsic."

8. Taylor (1986:45).

9. Taylor (1986:66).

10. Rolston (1994:168).

11. Schweitzer (1981:254).

12. Varner (1990:259).

13. Thompson (1990:153).

14. Some arguments for genic selectionism are discussed in chapter 6.

15. Goodpaster (1978:316).

16. Darwin (1981).

17. See Richards (1987:218–19) for a useful discussion of this view.

18. Rolston (1988) also offers a pluralistic account.

19. Taylor (1981:210).

20. Goodpaster (1978:320).

21. Goodpaster (1978:319).

22. Varner (1998:58).

23. Varner (1998:58).

24. See Passmore (1974:123) for a well-known expression of this concern.

25. Taylor (1986:45).

26. Taylor (1986:265).

27. Taylor (1986:266).

28. Taylor (1986:177–78).

29. At times Taylor appears to be introducing elements that might help him avoid this conclusion. For example, at one point he seems to give greater weight to moral agents (Taylor [1986:265]). Elsewhere, he claims that though sentience may not increase the value of an individual, it may add to ways in which that individual may be harmed (Taylor [1986:295]). These either do not help Taylor avoid the conclusion described above or, like Rolston's account, add higher values.

30. Rolston (1988:73).

31. For a survey of prospects for pluralism in ethics, see Becker (1992).

32. Varner quoting from Williams (1981:12–13).

33. Varner (1998:93).

34. Varner (1998:93).

35. Varner (1998:95).

5. A MORALLY SPECIALIZED ACCOUNT OF LIFE

1. For more detail on the following two definitions of life, see Emmeche (1994).

2. For readable and fascinating accounts of work on artificial life, see Emmeche (1994) and Levy (1993).

3. See White (1967).

4. See Keeton, Gould, and Gould (1993:chapter 19) for a description of conditions on the ancient Earth.

5. See de Duve (1995) for a helpful description of the early stages in the evolution of complex organisms.

6. Keeton, Gould, and Gould (1993:561–63).

7. Millikan (1989a), McGinn (1989), Papineau (1993).

8. Millikan (1986).

9. Fodor (1990).

10. Millikan (1989a), McGinn (1989), and Sterelny (1990) offer various solutions to the indeterminacy problem.

11. See the articles in Barkow, Cosmides, and Tooby (1992).

12. Taylor (1986:chapter 6) makes a similar point in distinguishing between basic and nonbasic interests. I consider it an advantage of the account I defend that it makes this distinction without introducing any new elements. Whether a goal is central or peripheral is simply a matter of how it affects other goals.

13. Though not necessarily all. An individual can have kin-related goals, satisfiable even in death. See chapter 7.

14. See chapter 7.

15. See Neander (1995).

16. See chapter 6 for more on this issue.

6. THE CONTENTS OF BIOPREFERENCES

1. See Millikan (1989a) for a discussion of the differences between simple and sophisticated representers.

2. See Lettvin, Maturana, McColloch, and Pitts (1959) for some of the original work on the frog's visual capacities. More recent discussions of the prey-catching capacities of frogs and toads can be found in Ewert (1987) and Ewert and Arbib (1989).

3. Fodor (1984).

4. Sterelny (1990:124).

5. Fodor (1990:18).

6. See Agar (1993) and Neander (1995) for contributions to this debate.

7. See Sober (1984:97–103) on what it means to say that a certain biological feature has been the target of natural selective forces.

8. Dawkins (1982:161).

9. Dawkins (1989:83).

10. Dawkins (1989:83).

11. Dawkins does recognize that even a piece of Xerox paper might be active to a degree. Its size and color will influence its probability of being copied.

12. Dawkins (1989:83).

13. See, for example, Gould (1987:77).

14. Dawkins (1989:77).

15. Dawkins (1989:23).

16. Dawkins (1989:200).

17. Dawkins (1989:200).

18. Dawkins (1989:210).

19. Dawkins (1989:206).

20. Dawkins (1989:21).

21. This is not to say that amoebas and the like do not represent difficult cases for the conventional biologist.

22. See the collection of articles in Kevles and Hood (1992) for a description of some of this research.

23. For one of the original papers on selfish DNA, see Doolittle and Sapienza (1980).

24. See, for example, Neander (1991b).

25. See, for example, Dennett (1991).

26. Singer (1993:126).

27. Singer (1993:126).

28. Singer (1993:127).

29. Sober (1990:102).

30. McGinn (1989:147).

31. Ng (1995:270).

32. Dretske (1988) is the best-known defense and elaboration of this view.

33. Nozick (1974:42).

34. See Unger (1990:chapter 9) for one such argument.

35. However, for a persuasive defense of captive breeding in zoos, see Tudge (1992).

36. Of course, we should not say the same of all domestic animals. Dogs, for example, have coexisted with humans for more than twelve thousand years. Intense selection has shaped their behavior and biopreferences to suit their domestic environs.

37. For just one attempt to demonstrate this conflict, see the argument in Sober (1986) that animal welfarists cannot be species preservationists.

7. Species and Ecosystems

1. Sober (1986:175).

2. Callicott (1980:320).

3. Norton (1987:168).

4. Callicott (1994:151).

5. Rolston (1994:174).

6. Taylor (1989) applies a similar distinction to the debate in political philosophy between Liberals and Communitarians.

7. Lovelock (1991:11).

8. Rolston (1988:75).

9. Johnson (1993:178).

10. Johnson (1993:180).

11. Dawkins (1989:chapter 14).

12. See Sober (1993:153–59) for a brief discussion of these species concepts.

13. Johnson (1993:178).

14. See King (1984:141–42) and Gaze (1994:43–47) for fuller accounts of the threats faced by the black stilt.

15. Researchers debate whether the main threat to the black stilt is predation or cross-species breeding. King (1984) suggests that black stilts will breed only with pied stilts when there are no other black stilts to be found. If this is so, then Mayr's definition might separate them. A pied stilt and a black stilt will breed only in an abnormal environment.

16. For a description of recent studies of Galapagos finches, see Weiner (1995).

17. See Wilson and MacArthur (1967).

18. Johnson (1993:216–17).

19. Johnson (1993:217).

20. White and Pickett (1985:7).

21. Reice (1994:427).

22. Reice (1994:427).

23. Reice (1994:428).

24. See, for example, Rolston (1988).

25. Sylvan and Bennett (1994).

26. Wynne-Edwards (1962).

27. Williams, of course, is a proponent of the gene rather than the individual as the unit of selection.

28. Williams (1966) and Dawkins (1989) have examples and explanations of biological altruism.

29. Wilson (1992:230).

30. See Gaze (1994:61–63). For a detailed description of the miraculous preservation effort, see Butler and Merton (1992).

31. Holmes (1994).

8. An Impossible Ethic?

1. Flanagan (1991:32).

2. Flanagan (1991:33).

3. Flanagan (1991:33).

4. Dennett (1995:498).

5. Williams and Smart (1973).

6. The worry that biocentric value will leave no room not only for what we would judge to be a good human life, but also for any human life whatsoever, is expressed in Passmore (1974).

7. Taylor (1986:312–13).

8. See, for example, Adams (1976).

9. Singer (1972), (1993:chapter 8).

10. For an attempt to tackle this type of problem, see Singer's response to Rolston in Jamieson (1999).

11. Eaton (1974:65–71).

12. Norton (1992) and Regan (1992).

13. The fact that one's motivational energies may already be exhausted by the need to look out for all humans certainly does not mean that there is no practical difference between the two accounts. We will not always place humans ahead of nonhumans. However, my point is that we do not increase the scale of burden.

14. Singer (1993:70–71).

Ackrill, J. L. 1981. *Aristotle the Philosopher*. Oxford: Oxford University Press.

Adams, R. 1976. "Motive Utilitarianism." *Journal of Philosophy* 73:467–81.

Agar, N. 1993. "What Do Frogs Really Believe?" *Australasian Journal of Philosophy* 71:1–12.

———. 1995. "Valuing Species and Valuing Individuals." *Environmental Ethics* 17:397–415.

———. 1996. "Teleology and Genes." *Biology and Philosophy* 11:289–300.

———. 1997. "Biocentric Ethics and the Concept of Life." *Ethics* 108:147–68.

Allaby, M. 1995. *Facing the Future: The Case for Science*. London: Bloomsbury.

Appleyard, B. 1992. *Understanding the Present*. London: Pan Books.

Barkow, J., Cosmides, L., and Tooby, J., eds. 1992. *The Adapted Mind: Evolutionary Psychology and the Generation of Culture*. New York: Oxford University Press.

Becker, L. 1992. "Places for Pluralism." *Ethics* 102:707–19.

Bigelow, J., and Pargetter, R. 1987. "Functions." *Journal of Philosophy* 84:181–96.

———. 1990. *Science and Necessity*. New York: Cambridge University Press.

Block, N. 1981. "Psychologism and Behaviorism." *Philosophical Review* 90:5–43.

———. 1991. "Troubles with Functionalism." In Rosenthal, D., ed. 1991. *The Nature of Mind*. Oxford: Oxford University Press.

Boorse, C. 1984. "Wright on Functions." In Sober, E., ed. 1984b. *Conceptual Issues in Evolutionary Biology*. Cambridge, Mass.: Bradford Books/MIT Press.

Boyd, R. 1988. "How to Be a Moral Realist." In Sayre-McCord, G., ed. 1988. *Essays on Moral Realism*. Cornell: Cornell University Press.

Brink, D. 1989. *Moral Realism and the Foundations of Ethics*. Cambridge: Cambridge University Press.

Budiansky, S. 1995. *Nature's Keepers: The New Science of Nature Management*. London: Weidenfeld and Nicholson.

Butler, D., and Merton, D. 1992. *The Black Robin: Saving the World's Most Endangered Bird*. Auckland, New Zealand: Oxford University Press.

Callicott, J. 1980. "Animal Liberation: A Triangular Affair." *Environmental Ethics* 2:324–45.

———. 1989. "Intrinsic Value, Quantum Theory, and Environmental Ethics." *Environmental Ethics* 7:257–75.

———. 1994. "The Conceptual Foundations of the Land Ethic." In Van DeVeer, D., and Pierce, C., eds. 1994. *The Environmental Ethics and Policy Book: Philosophy, Ecology, Economics.* Belmont, Calif.: Wadsworth Publishing.

Campbell, K. 1986. "Can Intuitive Psychology Survive the Growth of Neuroscience?" *Inquiry* 29:141–52.

Churchland, P. M. 1979. "Eliminative Materialism and the Propositional Attitudes." *Journal of Philosophy* 78:67–90.

Churchland, P. S. 1985. *Neurophilosophy: Toward a Unified Science of the Mind-Brain.* Cambridge, Mass.: Bradford Books/MIT Press.

Cummins, R. 1989. *Meaning and Mental Representation.* Cambridge, Mass.: Bradford Books/MIT Press.

Darwall, S., Gibbard, A., and Railton, P. 1992. "Towards Fin de Siecle Ethics." *The Philosophical Review* 101:115–90.

Darwin, C. 1981. *The Descent of Man and Selection in Relation to Sex.* Princeton: Princeton University Press.

Davidson, D. 1987. "Knowing One's Own Mind." *Proceedings and Addresses of the American Philosophical Association* 60:441–58.

Dawkins, M. 1990. "From an Animal's Point of View: Motivation, Fitness, and Animal Welfare." *Behavioral and Brain Sciences* 13:1–9.

———. 1993. *Through Our Eyes Only?: The Search for Animal Consciousness.* Oxford: W. H. Freeman.

Dawkins, R. 1982. "Replicators and Vehicles." In King's College Sociobiology Group, eds. *Current Problems in Sociobiology.* Cambridge: Cambridge University Press.

———. 1989. *The Extended Phenotype.* Oxford: Oxford University Press.

———. 1990. *The Selfish Gene.* Oxford: Oxford University Press.

De Duve, C. 1995. *Vital Dust: Life As a Cosmic Imperative.* New York: Basic Books/HarperCollins.

DeGrazia, D. 1996. *Taking Animals Seriously: Mental Life and Moral Status.* Cambridge: Cambridge University Press.

Dennett, D. 1991. *Consciousness Explained.* Boston: Little, Brown and Company.

———. 1995. *Darwin's Dangerous Idea: Evolution and the Meanings of Life.* London: Penguin.

Devitt, M., and Sterelny, K. 1987. *Language and Reality: An Introduction to the Philosophy of Language.* Oxford: Blackwell.

Doolittle, W., and Sapienza, C. 1980. "Selfish Genes, the Phenotypic Paradigm, and Genome Evolution." *Nature* 284:601–3.

Dretske, F. 1981. *Knowledge and the Flow of Information*. Oxford: Blackwell.

———. 1988. *Explaining Behavior*. Cambridge, Mass.: Bradford Books/MIT Press.

Dupre, J. 1993. *The Disorder of Things: Metaphysical Foundations of the Disunity of Science*. Cambridge, Mass.: Harvard University Press.

Eaton, R. 1974. *The Cheetah: The Biology, Ecology and Behavior of an Endangered Species*. New York: Van Nostrand Reinhold Company.

Editorial commentary on Schull, J. 1990. "Are Species Intelligent?" *Behavioral and Brain Sciences* 13:94.

Elliot, R. 1997. *Faking Nature: The Ethics of Environmental Restoration*. London: Routledge and Kegan Paul.

Elton, C. 1927. *Animal Ecology*. London: Sidgwick and Jackson.

Emmeche, C. 1994. *The Garden in the Machine: The Emerging Science of Artificial Life*. Princeton: Princeton University Press.

Erlich, A., and Erlich, P. 1994. Extract from *The Population Explosion*. Reprinted in Gruen, L., and Jamieson, D. 1994. *Reflecting on Nature: Readings in Environmental Philosophy*. Oxford: Oxford University Press:309–20.

Ewert, J-P. 1987. "Neuroethology of Releasing Mechanisms: Prey-catching in Toads." *Behaviour and Brain Sciences* 10:337–405.

Ewert, J-P., and Arbib, M., eds. 1989. *Visuomotor Coordination: Amphibians, Comparisons, Models, and Robots*. New York: Plenum Press.

Feinberg, J. 1974. "The Rights of Animals and Unborn Generations." In Blackstone, W., ed. *Philosophy and Environmental Crisis*. Athens: University of Georgia Press.

Flanagan, O. 1991. *Varieties of Moral Personality: Ethics and Psychological Realism*. Cambridge, Mass.: Harvard University Press.

Fodor, J. 1968. *Psychological Explanation: An Introduction to the Philosophy of Psychology*. New York: Random House.

———. 1983. "The Present Status of the Innateness Controversy." In his *RePresentations: Philosophical Essays on the Foundations of Cognitive Science*. Cambridge, Mass.: Bradford Books/MIT Press.

———. 1984. "Semantics, Wisconsin Style." *Synthese* 59:231–50.

———. 1986. "Why Paramecia Don't Have Mental Representations." In French, P., Uehling, T., and Wettstein, H., eds. *Midwest Studies in Philosophy* 10. Minneapolis: University of Minnesota Press.

———. 1987. *Psychosemantics: The Problem of Meaning in the Philosophy of Mind*. Cambridge, Mass.: Bradford Books/MIT Press.

———. 1990. *A Theory of Content and Other Essays*. Cambridge, Mass.: Bradford Books/MIT Press.

Frey, R. G. 1980. *Interests and Rights: The Case Against Animals*. Oxford: Clarendon Press.

Gaze, P. 1994. *Rare and Endangered New Zealand Birds: Conservation and Management.* Christchurch: Canterbury University Press.

Gibson, E., McKay, D., Thomas Keprta, K., and Romanek, C. 1997. "The Case for Relic Life on Mars." *Scientific American* 277 (December):36–41.

Godfrey-Smith, P. 1994. "A Modern History Theory of Function." *Nous* 28:344–62.

Goodpaster, K. 1978. "On Being Morally Considerable." *Journal of Philosophy* 75:308–25.

Gould, S. 1987. *The Panda's Thumb: Reflections on Natural History.* London: Penguin.

Gould, S., and Lewontin, R. 1984. "The Spandrels of San Marco and the Panglossian Paradigm: A Critique of the Adaptationist Programme." In Sober, E., ed. *Conceptual Issues in Evolutionary Biology.* Cambridge, Mass.: Bradford Books/MIT Press.

Grumbine, E. 1992. *Ghost Bears: Exploring the Biodiversity Crisis.* Washington, D.C.: Island Press.

Hacking, I. 1995. *Rewriting the Soul: Multiple Personality and the Sciences of Memory.* Princeton: Princeton University Press.

Hall, C. 1990. "Sanctioning Resource Depletion: Economic Development and Neo-Classical Economics." *The Ecologist* 20(3).

Hargrove, E. 1992. "Weak Anthropocentric Intrinsic Value." *Monist* 75:181–207.

Holland, J., Holyoak, K., Nisbett, J., and Thagard, P. 1989. *Induction: Processes of Inference, Learning, and Discovery.* Cambridge, Mass.: Bradford Books/MIT Press.

Holmes, B. 1994. "Second Chance for the Tuatara." *New Scientist* (25 Nov. 1995):8.

Hoyle, F., and Wickramasinghe, N. C. 1993. *Our Place in the Cosmos: The Unfinished Revolution.* London: J. M. Dent.

Hume, D. 1978. *Treatise on Human Nature.* Selby-Bigge, L., ed. Oxford: Clarendon Press.

Jackson, F. 1991. "What Mary Didn't Know." In Rosenthal, D., ed. 1991. *The Nature of Mind.* Oxford: Oxford University Press.

Jackson, F., and Pettit, P. 1989. "In Defence of Folk Psychology." *Philosophical Studies* 59:31–54.

Jakosky, B. 1996. "Warm Havens for Life on Mars." *New Scientist* (4 May 1996) 2028:38–42.

Jamieson, D., ed. 1999. *Singer and His Critics.* Oxford: Basil Blackwell.

Janzen, D. 1977. "What Are Dandelions and Aphids?" *American Naturalist* 111:586–89.

Johnson, L. 1992. "Toward the Moral Considerability of Species and Ecosystems." *Environmental Ethics* 14:145–57.

———. 1993. *A Morally Deep World: An Essay on the Moral Significance and Environmental Ethics.* Cambridge: Cambridge University Press.

Keeton, W., Gould, J., and Gould, G. 1993. *Biological Science.* New York: W. W. Norton.

Kekes, J. 1993. *The Morality of Pluralism.* Princeton: Princeton University Press.

Kevles, D., and Hood, L. 1992. *The Code of Codes: Scientific and Social Issues in the Human Genome Project.* Cambridge, Mass.: Harvard University Press.

King, C. 1984. *Immigrant Killers: Introduced Predators and the Conservation of Birds in New Zealand.* Auckland: Oxford University Press.

Korsgaard, C. 1996. *Creating the Kingdom of Ends.* Cambridge: Cambridge University Press.

Kripke, S. 1980. *Naming and Necessity.* Cambridge, Mass.: Harvard University Press.

Leopold, A. 1949. *A Sand County Almanac.* New York: Oxford University Press.

Lettvin, J., Maturana, H., McColloch, W., and Pitts, W. 1959. "What the Frog's Eye Tells the Frog's Brain." *Proceedings of the Institute of Radio Engineers:*1940–57.

Levy, S. 1993. *Artificial Life: The Quest for a New Creation.* London: Penguin.

Lewis, D. 1994. "Reduction of Mind." In Guttenplan, S., ed. *Companion to the Philosophy of Mind.* Oxford: Blackwells.

Lewontin, R. 1992. *Biology As Ideology: The Doctrine of DNA.* New York: Harper Perennial.

Lovelock, J. 1988. In Wilson, E. O., and Peter, F. M., eds. 1988. *Biodiversity.* Washington, D.C.: National Academy Press.

———. 1991. *Healing Gaia: Practical Medicine for the Planet.* New York: Harmony Books.

Lycan, W. 1987. *Consciousness.* Cambridge, Mass.: Bradford Books/MIT Press.

MacArthur, R., and Wilson, E. O. 1967. *The Theory of Island Biogeography.* Princeton: Princeton University Press.

McGinn, C. 1989. *Mental Content.* Oxford: Blackwell.

Maturana, H., and Varela, F. 1980. *Autopoiesis and Cognition.* Dordrecht: Reidel.

Mayr, E. 1963. *Animal Species and Evolution.* Cambridge, Mass.: Harvard University Press.

Medawar, P. 1957. *The Uniqueness of the Individual.* London: Methuen.

Midgely, M. 1979. "Gene-juggling." *Philosophy* 54:439–58.

Mill, J. S. 1861. *Utilitarianism.* London: John Fraser.

Millikan, R. 1986. "Thoughts without Laws: Cognitive Science with Content." *Philosophical Review* 95:47–80.

———. 1989a. "Biosemantics." *Journal of Philosophy* 86:281–97.

———. 1989b. "In Defence of Proper Functions." *Philosophy of Science* 56:288–302.

Moore, G. E. 1922. *Philosophical Studies.* London: Routledge and Kegan Paul.

Nagel, T. 1974. "What Is It Like to a Bat?" *Philosophical Review* 83:435–50.

Naess, A., and Rothenberg, D. 1989. *Ecology, Community and Lifestyle: Ecosophy T.* Cambridge: Cambridge University Press.

Nash, R. 1990. *The Rights of Nature: A History of Environmental Ethics.* Australia: Primavera Press.

Neander, K. 1991a. "The Teleological Notion of 'Function.' " *Australasian Journal of Philosophy* 69:454–68.

———. 1991b. "Functions As Selected Effects: The Conceptual Analyst's Defence." *Philosophy of Science* 58:168–84.

———. 1995. "Misrepresenting and Malfunctioning." *Philosophical Studies* 79/2:109–41.

Nesse, R., and Williams, G. 1995. *Evolution and Healing: The New Science of Darwinian Healing.* London: Weidenfeld and Nicolson.

Ng, Y-K. 1995. "Towards Welfare Biology: Evolutionary Economics of Animal Consciousness and Suffering." *Biology and Philosophy* 10:255–85.

Norton, B. 1987. *Why Preserve Natural Variety?* Princeton: Princeton University Press.

———. 1992. "Epistemology and Environmental Values." *Monist* 75:208–26.

Nozick, R. 1974. *Anarchy, State and Utopia.* New York: Basic Books.

O'Neill, J. 1992. "The Varieties of Intrinsic Value." *Monist* 75:119–37.

Papineau, D. 1993. *Philosophical Naturalism.* Oxford: Basil Blackwell.

Passmore, J. 1974. *Man's Responsibility for Nature.* London: Duckworth.

Pigden, C. 1991. "Naturalism." In Singer, P., ed. *A Companion to Ethics.* Oxford: Blackwell.

Putnam, H. 1979. "The Meaning of 'Meaning.' " In Gunderson, K., ed. *Language, Mind, and Knowledge. Vol 7, Minnesota Studies in the Philosophy of Science.* Minneapolis: University of Minnesota Press.

Rachels, J. 1990. *Created from Animals: The Moral Implications of Darwinism.* Oxford: Oxford University Press.

Railton, P. 1986. "Moral Realism." *Philosophical Review* 95:163–207.

Regan, T. 1983. *The Case for Animal Rights.* Berkeley: University of California Press.

———. 1992. "Does Environmental Ethics Rest on a Mistake?" *Monist* 75:161–82.

Reice, S. 1994. "Nonequilibrium Determinants of Biological Community Structure." *American Scientist* 82(5):424–35.

Richards, R. 1987. *Darwin and the Emergence of Evolutionary Theories of Mind and Behavior.* Chicago: University of Chicago Press.

Rollin, B. 1990. *The Unheeded Cry: Animal Consciousness, Animal Pain and Science.* Oxford: Oxford University Press.

Rolston, H., III. *Conserving Natural Value.* New York: Columbia University Press.

———. 1988. *Environmental Ethics: Values in and Duties to the Natural World.* Philadelphia: Temple University Press.

Ruse, M. 1995. *Evolutionary Naturalism.* London: Routledge.

Safina, C. 1995. "The World's Most Imperiled Fish." *Scientific American* (Nov. 1995):30–43.

Schrader-Frechette, K. 1995. "Practical Ecology and Foundations for Environmental Ethics." *Journal of Philosophy* 92:621–35.

Schull, J. 1990. "Are Species Intelligent?" *Behavioral and Brain Sciences* 13:63–74.

Schweitzer, A. 1953. *Out of My Life and Thought: An Autobiography.* New York: New American Library.

———. 1981. *Philosophy of Civilization.* Tallahassee: University Presses of Florida.

Sheldrake, R. 1981. *A New Science of Life: The Hypothesis of Formative Causation.* Los Angeles: J. P. Tarcher.

Singer, P. 1972. "Famine, Affluence and Morality." *Philosophy and Public Affairs* 1:229–43.

———. 1979. "Not for Humans Only: The Place of Nonhumans in Environmental Issues." In Goodpaster, K., and Sayre, K., eds. *Ethics and Problems of the 21st Century.* Notre Dame: University of Notre Dame Press.

———. 1993. *Practical Ethics: Second Edition.* Cambridge: Cambridge University Press.

Singer, P., and Cavalieri, P., eds. 1993. *The Great Ape Project: Equality beyond Humanity.* London: Fourth Estate.

Smith, M. 1991. "Moral Realism." In Singer, P., ed. *A Companion to Ethics.* Oxford: Blackwell.

Sober, E. 1984. *The Nature of Selection.* Cambridge, Mass.: Bradford Books/MIT Press.

———. 1986. "Philosophical Problems for Environmentalism." In Norton, B., ed. *The Preservation of Species.* Princeton: Princeton University Press.

———. 1990. "Putting the Function Back into Functionalism." In Lycan, W., ed. *Mind and Cognition: A Reader.* Oxford: Basil Blackwell.

———. 1993. *Philosophy of Biology.* New York: Oxford University Press.

Sterelny, K. 1990. *The Representational Theory of the Mind: An Introduction.* Oxford: Blackwell.

———. 1995. Critical Notice on Barkow, Cosmides, and Tooby (1992). *Biology and Philosophy* 10:365–80.

Sterelny, K., and Kitcher, P. 1988. "The Return of the Gene." *Journal of Philosophy* 85:339–61.

Stich, S. 1983. *From Folk Psychology to Cognitive Science.* Cambridge, Mass.: Bradford Books/MIT Press.

———. 1992. "What Is a Theory of Mental Representation?" *Mind* 101:241–61.

Sylvan, R., and Bennett, D. 1994. *The Greening of Ethics.* Cambridge: White Horse Press.

Taylor, C. 1989. " 'Cross-purposes': The Liberal-Communitarian Debate." In Roseblum, N., ed. *Liberalism and the Moral Life.* Cambridge: Harvard University Press.

Taylor, P. 1981. "The Ethics of Respect for Nature." *Environmental Ethics* 3:197–218.

———. 1984. "Are Humans Superior to Animals and Plants?" *Environmental Ethics* 6:149–60.

———. 1986. *Respect for Nature: A Theory of Environmental Ethics.* Princeton: Princeton University Press.

Thompson, J. 1990. "A Refutation of Environmental Ethics." *Environmental Ethics* 12:147–60.

Thompson, P., ed. 1995. *Issues in Evolutionary Ethics*. Albany, N.Y.: State University of New York Press.

Tooley, M. 1972. "Abortion and Infanticide." *Philosophy and Public Affairs* 2:37–65.

Tudge, C. 1992. *Last Animals at the Zoo: How Mass Extinction Can Be Stopped*. Oxford: Oxford University Press.

Turing, A. 1950. "Computing Machinery and Intelligence." In *Mind* 59:433–60.

Unger, P. 1990. *Identity, Consciousness and Value*. New York: Oxford University Press.

———. 1996. *Living High and Letting Die: Our Illusion of Innocence*. New York: Oxford University Press.

Van Valen, L. 1976. "Ecological Species, Multispecies, and Oaks." *Taxon* 25:233–39.

Varner, G. 1990. "Biological Functions and Biological Interests." *Southern Journal of Philosophy* 27:251–70.

———. 1998. *In Nature's Interests?: Interests, Animal Rights and Environmental Ethics*. Oxford: Oxford University Press.

Weiner, J. 1995. *The Beak of the Finch: A Story of Evolution in Our Time*. New York: Random House.

White, Lynn, Jr. 1967. "The Historical Roots of Our Ecologic Crisis." *Science* 155:1203–7.

White, P., and Pickett, S., eds. 1985. *The Ecology of Natural Disturbance and Patch Dynamics*. New York: Academic Press.

Williams, B. 1991. "Consequentialism and Integrity." In Scheffler, S., ed. 1991. *Consequentialism and Its Critics*. Oxford: Oxford University Press.

Williams, B., and Smart, J. 1973. *Utilitarianism: For and Against*. Cambridge: Cambridge University Press.

Williams, G. 1966. *Adaptation and Natural Selection*. Princeton: Princeton University Press.

Wilson, E. O. 1979. *On Human Nature*. Cambridge, Mass.: Harvard University Press.

———. 1984. *Biophilia*. Cambridge, Mass.: Harvard University Press.

———. 1992. *The Diversity of Life*. London: Penguin Press.

———. 1994a. "The Little Things that Run the World." In Van DeVeer, D., and Pierce, C., eds. 1994. *The Environmental Ethics and Policy Book: Philosophy, Ecology, Economics*. Belmont, Calif.: Wadsworth Publishing.

Wilson, E. O., and Kellert, S., eds. 1993. *The Biophilia Hypothesis*. Washington, D.C.: Island Press.

Wilson, E. O., and MacArthur, R. H. 1967. *The Theory of Island Biogeography*. Princeton: Princeton University Press.

Wright, L. 1973. "Functions." *Philosophical Review* 85:70–86.

Wynne-Edwards, V. C. 1962. *Animal Dispersion in Relation to Social Behaviour*. Edinburgh: Oliver and Boyd.

Index

Printed in the USA
CPSIA information can be obtained
at www.ICGtesting.com
JSHW021436221024
72172JS00002B/27